Springer Theses

Recognizing Outstanding Ph.D. Research

For further volumes:
http://www.springer.com/series/8790

Aims and Scope

The series "Springer Theses" brings together a selection of the very best Ph.D. theses from around the world and across the physical sciences. Nominated and endorsed by two recognized specialists, each published volume has been selected for its scientific excellence and the high impact of its contents for the pertinent field of research. For greater accessibility to non-specialists, the published versions include an extended introduction, as well as a foreword by the student's supervisor explaining the special relevance of the work for the field. As a whole, the series will provide a valuable resource both for newcomers to the research fields described, and for other scientists seeking detailed background information on special questions. Finally, it provides an accredited documentation of the valuable contributions made by today's younger generation of scientists.

Theses are accepted into the series by invited nomination only and must fulfill all of the following criteria

- They must be written in good English.
- The topic should fall within the confines of Chemistry, Physics, Earth Sciences, Engineering and related interdisciplinary fields such as Materials, Nanoscience, Chemical Engineering, Complex Systems and Biophysics.
- The work reported in the thesis must represent a significant scientific advance.
- If the thesis includes previously published material, permission to reproduce this must be gained from the respective copyright holder.
- They must have been examined and passed during the 12 months prior to nomination.
- Each thesis should include a foreword by the supervisor outlining the significance of its content.
- The theses should have a clearly defined structure including an introduction accessible to scientists not expert in that particular field.

Eric McCalla

Consequences of Combinatorial Studies of Positive Electrodes for Li-ion Batteries

Springer

Eric McCalla
Physics and Atmospheric Sciences Dept.
Dalhousie University
Halifax, Nova Scotia
Canada

ISSN 2190-5053 ISSN 2190-5061 (electronic)
ISBN 978-3-319-38317-0 ISBN 978-3-319-05849-8 (eBook)
DOI 10.1007/978-3-319-05849-8
Springer Cham Heidelberg New York Dordrecht London

Printed on acid-free paper

Springer is part of Springer Science+Business Media (www.springer.com)

A learned blockhead is a greater blockhead than an ignorant one.

–Benjamin Franklin

I dedicate this thesis to my mother, Joanne McCalla, who was my first science teacher and helped me become a spectacular blockhead.

Foreword for Eric McCalla's Thesis-prepared by Jeff Dahn

Given the large number of attractive and useful materials in the Li–Ni–Mn–O system and given the confusion about the structure and properties of many of them, Eric McCalla decided to undertake a complete determination of the Li–Ni–Mn–O phase diagram as the major portion of his Ph.D. thesis. Eric adopted combinatorial and high throughput materials science methods for the synthesis of these materials and used X-ray diffraction as the main characterization tool. Eric learned how to control lithium loss during heating of the small combinatorial samples and studied the effects of oxygen partial pressure, heating temperature, and cooling rate on the obtained materials.

The phase diagrams presented in this thesis have all been determined with great care. Analysis of multiphase regions using the lever rule gave phase boundaries that agree very well with direct determinations of phase boundaries by inspection of X-ray patterns for single phase or hints of other phases. The dramatic effect of cooling rate on the three phase regions in the phase diagrams was a huge surprise. In addition, the identification of a solid solution phase, $Li[Li_{1/3-x}\square_{x/2}Ni_{x/2}Mn_{2/3}]O_2$, where Ni and a vacancy ($\square$) replace two Li atoms in $Li[Li_{1/3}Mn_{2/3}]O_2$, showed that lithium-deficient layered materials incorporate vacancies, not Mn^{3+}.

The thesis shows that "layered-layered" nanocomposites and the "layered-layered-spinel" three phase composites do exist in certain regions of the phase diagram. However, these materials do not incorporate $LiNi_{1/2}Mn_{1/2}O_2$ and Li_2MnO_3 as the two layered components. Instead, the Mn-rich component contains some Ni and vacancies and the Ni-rich component is not $LiNi_{1/2}Mn_{1/2}O_2$. Furthermore, Eric shows in the thesis that materials which are "layered-layered" nanocomposites yield inferior electrochemical performance to those which are single-phase solid solutions. The thesis also explains why rock-salt structure so-called "impurity" phases occur in many samples in the literature and that these rock-salt phases are not impurities but are expected based on the Li–Ni–Mn–O phase diagram. There are many more gems in the thesis that the reader can discover.

The thesis was examined and approved by Dr. Shirley Meng of the University of California at San Diego. Dr. Meng was impressed by the gap that Eric's thesis helped fill in the cathode materials research community. Meng believes that Eric's methodology was unique and the information Eric presented in his thesis will be critical to researchers in the future.

It was my pleasure to be involved in this work as Eric's supervisor and to be able to interact with him on a daily basis. Eric was an excellent student who solved many puzzles developing the results in this thesis. The methodology developed by Eric will be used to complete future studies on the entire Li–Ni–Mn–Co–O system.

11 Feb, 2014 Jeff Dahn
Dalhousie University

Preface

The improvement of volumetric energy density remains a key area of research to optimize Li-ion batteries for applications such as extending the range of electric vehicles. There is still improvement to be made in the energy density in the positive electrode materials. The current thesis deals with determining the phase diagrams of the Li–Mn–Ni–O and Li–Co–Mn–O systems in order to better understand the structures and the electrochemistry of these materials. The phase diagrams were made through careful analysis of hundreds of X-ray diffraction patterns taken of milligram-scale combinatorial samples. A number of bulk samples were also investigated.

The Li–Mn–Ni–O system is of particular interest as avoiding cobalt lowers the cost of the material. However, this system is very complex: there are two large solid-solution regions separated by three two-phase regions as well as two three-phase regions. Comparing quenched and slow cooled samples shows that the system transforms dramatically when cooled at rates typically used to make commercial materials. The consequences of these results are that much of the system must be avoided in order to guarantee that the materials remain single phase during cooling. This work should therefore impact significantly researchers working on composite electrodes.

Two new structures were found. The first was Li-Ni-Mn oxide rocksalt structures with vacancies and ordering of manganese which were previously mistakenly identified as $Li_x Ni_{2-x}O_2$. The other new structure was a layered oxide with metal site vacancies allowing manganese to order on two $\sqrt{3} \times \sqrt{3}$ superlattices. The electrochemistry of both these materials is presented here.

Finally, the region where layered-layered composites form during cooling has been determined. These materials were long looked for along the composition line from $Li_2 MnO_3$ to $LiNi_{0.5} Mn_{0.5} O_2$ and the most significant consequence of the actual locations of the end-members is that one of the structures contains a high concentration of nickel on the lithium layer. Layered-layered nano-composites formed in this system are therefore not ideal positive electrode materials and it will be demonstrated that single-phase layered materials lead to better electrochemistry.

Preface

Acknowledgments

I would like to thank my supervisor, Dr. Jeff Dahn, for guidance, encouragement, and his generous financial support. I have had many interesting and challenging discussions with Dr. Dahn over the last three years and working in his lab has given me access to state of the art research equipment and opportunities to contact world class researchers in industrial and academic fields. I am extremely grateful for the opportunity to have done my doctorate in his research group.

Thanks also go out to the present and former staff in the Department of Physics and Atmospheric Science for their assistance during my time. Thank you to Anne Murphy, Barbara Gauvin, Tanya Timmins, Jennifer Currie, and Heather Anne Jennex in the physics office. I would also like to thank Andy George, Kevin Borgel, Dan Chevalier, and Simon Trussler for their help and advice.

I would like to thank all members of the Dahnlab who, over the past few years, have made this time both very enjoyable and sufficiently insane. A number of students were directly involved in this work, including Aaron Rowe, Colby Brown, Cassandra Lowartz, Cody Watson, John Camardese, Paul Duchesne, and Ramesh Shunmugasundaram and I would to thank them all for working and arguing with me. I would also like to thank Jennifer Romero, Robbie Sanderson, David Stevens, and Trevor Byrne for a lot of help with knowing how to use and fix a variety of equipment. I also thank all the coopians for much needed distractions and useful conversations! Thanks also go to Laura Downie who ensured I survived the writing of this thesis.

I am, as always, thankful to my family and friends who have supported me throughout.

Halifax, December 2013 Eric McCalla

Contents

List of Symbols and Abbreviations Used

2θ	Scattering angle
$2\theta_k$	Scattering angle for the center of the k-th X-ray diffraction peak
$\alpha,\ \beta,\gamma$	Angles defined in the Li-Mn-Ni-O Gibbs triangle
β_k	Integral breadth of the k-th X-ray diffraction peak
β_T	Constant used to set the temperature scale in Monte Carlo simulations
δ	Ratio of an atom's charge to its oxidation state, used in Monte Carlo simulations
ε_0	Permittivity of free space
η	Lorentzian component in the pseudo-Voigt function
λ	Wavelength
σ	Standard deviation
μ	Linear X-ray absorption coefficient
$\sqrt{<e>^2}$	Root-mean square micro-strain
AA	Atomic absorption spectroscopy
C	Lorentzian, or Cauchy, function
D	Particle diameter
F_k	Sample scattering X-ray diffraction intensity for the k-th peak
FWHM	Full width at half maximum
G	Gaussian function
H_k	FWHM of k-th X-ray diffraction peak
ICP	Inductively coupled plasma optical emission spectroscopy
I_k	Calculated X-ray diffraction intensity for the k-th peak
I_k^o	Integrated peak intensity for the k-th X-ray diffraction peak
IRC	Irreversible capacity
k_B	Boltzmann constant
L	Average crystallite size
M	XRD machine peak broadening function
JCPDS	Joint Committee on Powder Diffraction Standards
n_i	Oxidation state of i-th atom in Monte Carlo simulations
Ni_{Li}	Fraction of sites on the lithium layers occupied by nickel
NN	Nearest neighbor
SEM	Scanning electron microscopy

T	Temperature
TM	Transition metal
TGA	Thermo-gravimetric analysis
XRD	X-ray diffraction
XANES	X-ray absorption near-edge structure

List of Figures

List of Tables

About the Author

Eric McCalla was born and raised in Quebec city. He completed his undergraduate degree at Mount Allison University in 2000 before doing a Masters in Solid State Physics at McGill University. Following the completion of the Master's in 2002, he become a computer science teacher in Kuujjuaraapik, an Inuit community in Northern Quebec. After two years of teaching, he decided to become a teacher and so followed the Bachelor of Education program at the Memorial University of Newfoundland. He then worked as a high school mathematics and science teacher in a small high school in Sept-Iles, Quebec. On a sabbatical, after biking across Canada and needing distraction, he decided to return to graduate school in 2010. The results of the following three years at Dalhousie University working in the group of Jeff Dahn are the subject of this book. He now works as a Post-Doctoral Fellow at the Collège-de-France in Paris in the group of Jean-Marie Tarascon.

Chapter 1
Introduction

1.1 Motivation: Li–Co–Mn–Ni Oxide Materials

Improving the energy density of lithium (Li)-ion batteries remains important for a number of applications. In particular, high energy densities are required to extend the range of electric vehicles and minimize the space and mass of the battery pack. In the ongoing search for means to increase the energy of Li-ion batteries, the discovery of new positive electrode materials is of critical importance. The positive electrode is synthesized with lithium in it, some of which is removed during charging of the cell along with electrons that travel through the external circuit. The lithium is then reinserted into the material while the battery is discharged, which is accompanied by electrons traveling through the external circuit to do work. There are many challenges with respect to finding better positive electrode materials, the primary of which are increasing the volumetric energy density, lowering the cost, and improving the safety. Though safety is a very important issue with respect to Li-ion batteries, the focus of this thesis will be to maximize energy while trying to minimize cost.

Table 1.1 shows the volumetric energy density of the most competitive positive electrode materials, calculated from data from Ref. [1]. The most common commercially used positive electrode, $LiCoO_2$, has an energy density of 3.05 Wh/cm^3 while the material with the highest energy density, $Li[Li_{1/9}Ni_{1/3}Mn_{5/9}]O_2$, is a lithium-rich layered oxide material that has roughly 25 % more energy than $LiCoO_2$. The lithium-rich layered structures are very promising as possible next-generation high energy positive electrodes. They exist over wide composition ranges within the Li–Co–Mn–Ni–O system [2, 3] and will be discussed in detail throughout this thesis. Despite the potentially very high energy density, there remain challenges with respect to these materials and this will be discussed throughout the thesis. The primary motivation for studying the Li–Co–Mn–Ni–O system is, therefore, that many promising materials have already been found in this system. For example, all the materials listed in Table 1.1 lie within this system except for the last two, one of which is $LiFePO_4$ with the lowest energy density of those listed. It must be emphasized here that the primary objective here is maximizing volumetric energy density. For applications where gravimetric energy density and particularly power are more critical materials such as $LiFePO_4$ are of greater importance. Table 1.1, therefore, demonstrates why

E. McCalla, *Consequences of Combinatorial Studies of Positive Electrodes for Li-ion Batteries,* Springer Theses, DOI 10.1007/978-3-319-05849-8_1,

Table 1.1 The median voltage, specific capacity, crystallographic density and volumetric energy density of some oxide materials that are either commercially available or promising candidates for next generation Li-ion batteries

Material	Specific capacity (mAh/g)	Median voltage (V)	Crystallographic density (g/cm³)[a]	Volumetric energy (Wh/cm³)
$LiCoO_2$	155	3.9	5.05	3.05
$LiNi_{1-x-y}Mn_xCo_yO_2$	140–180	3.8	4.77 ($x = y = 1/3$)	2.54–3.26
$LiMn_2O_4$	100–120	4.05	4.29	1.74–2.08
$Li[Li_{1/9}Ni_{1/3}Mn_{5/9}]O_2$	240	3.8	4.45	4.06
$LiNi_{0.5}Mn_{1.5}O_4$	130	4.6	4.4	2.63
$LiFePO_4$	160	3.45	3.60	1.99
$LiNi_{0.8}Co_{0.15}Al_{0.05}O_2$	200	3.73	4.75	3.54

[a] The crystallographic densities were obtained from either the JCPDS database or from Refs. [2, 4, 5].

looking for materials with maximum energy in the Li–Co–Mn–Ni–O system is warranted for applications where the objective is to maximize energy within a limited volume. A particular effort must be made to minimize the cobalt content to lower the cost of the material.

Much work has already been performed on materials in the Li–Co–Mn–Ni–O system as potential positive electrode materials. Typical studies map out composition lines, and as a result, a number of useful single phase systems were discovered such as $Li_{1+x}Mn_{2-x}O_4$ [6–8] and $Li[Ni_xMn_xCo_{1-2x}]O_2$ [5, 9]. The current project is limited to two subsets of this system: Li–Mn–Ni–O and Li–Co–Mn–O. The objective is to map these two systems out in their entirety, which has never been done before. Carey and Dahn demonstrated that a combinatorial solution-based approach can be used to synthesize oxide materials by making Li–Ni–Mn–O spinel materials [10]. The main goal of this thesis is to adapt this method for use over the two systems of interest.

Each of the Li–Mn–Ni–O and Li–Co–Mn–O systems can be viewed as Gibbs triangles, though this implies plotting compositions based on metallic fractions only. As such, these systems will be referred to as "pseudo-ternary" because the oxygen content of the samples was not controlled and reached equilibrium concentrations as the samples were synthesized. However, nonquenched samples only reached near-equilibrium since the oxygen content for slow cooled samples is affected by the kinetics during cooling. Given that phase transformations occur during cooling [11], as will be discussed extensively throughout the thesis, the pseudo-ternary phase diagrams presented here are in fact phase stabilities, or metastabilities in the case of the slow cooled system where equilibrium conditions are never reached [12]. The phase diagrams typically shown for such systems prior to this work are far from complete with no studies looking at the materials over all composition ranges. This has severely limited the extent to which single-phase regions have been explored and has made studies of composite electrodes particularly difficult since the phases involved in the coexistence regions have not been precisely determined [13, 14].

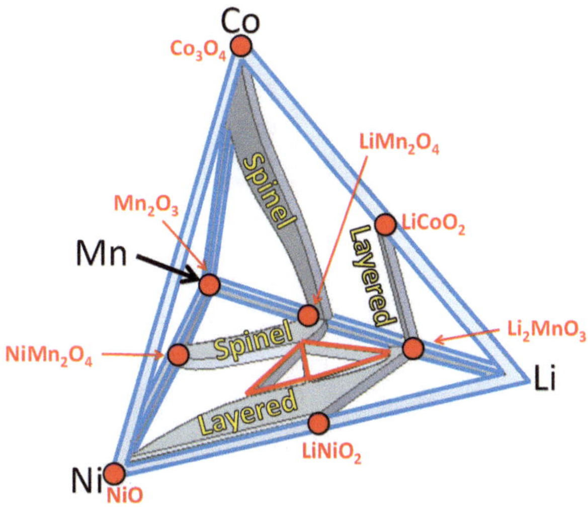

Fig. 1.1 The Li–Co–Mn–Ni–O pseudo-quaternary system of extreme interest for potential positive electrode materials. The *blue lines* indicate the axes, while the *red dots* represent a few of the single-phase materials known prior to the current study. The single-phase regions along the Li–Co–Mn and Li–Mn–Ni faces are shown and are the results of this project for samples quenched from 800 °C and the *red lines* bound three-phase regions

This thesis is, therefore, part of a project that has an ultimate goal of examining the entire Li–Co–Mn–Ni oxide pseudo-quaternary phase diagram and should be of significant interest to the Li-ion research community. Figure 1.1 shows the Gibbs pyramid representing the pseudo-quaternary system. The Li–Co–Mn and the Li–Ni–Mn oxide systems studied here are faces on the pyramid. Some of the results from this project for quenched samples have been included in Fig. 1.1.

The rest of this introduction deals primarily with the state of knowledge prior to the current project, with a particular emphasis on points of confusion in the literature where the phase diagrams developed here allow for a better understanding of previously published results. Throughout the introduction, these points of confusion will be numbered in bold roman numerals, i.e. **I**, **II**, **III** ... , and these will be revisited in Chap. 10.

1.2 Layered and Spinel Structures

Both faces of interest in Fig. 1.1 contain two large single-phase regions of high promise as potential positive electrode materials: the layered and spinel phases. There are materials with either of these structures that are commercially relevant and each will be discussed in depth in this thesis. The layered materials can be viewed as ordered rocksalt structures with parallel sheets of atoms where each sheet is made up of hexagonal arrays. These layers alternate following the pattern: lithium, oxygen, transition metal (TM), oxygen, lithium ... All lithium atoms are in octahedral sites in these structures (six nearest neighbor oxygens forming an octahedron). Figure 1.2 shows the structures of the layered and spinel phases. Examples of layered metal oxides that are currently used commercially are: $Li[Co]O_2$ and $Li[Ni_{1/3}Mn_{1/3}Co_{1/3}]O_2$,

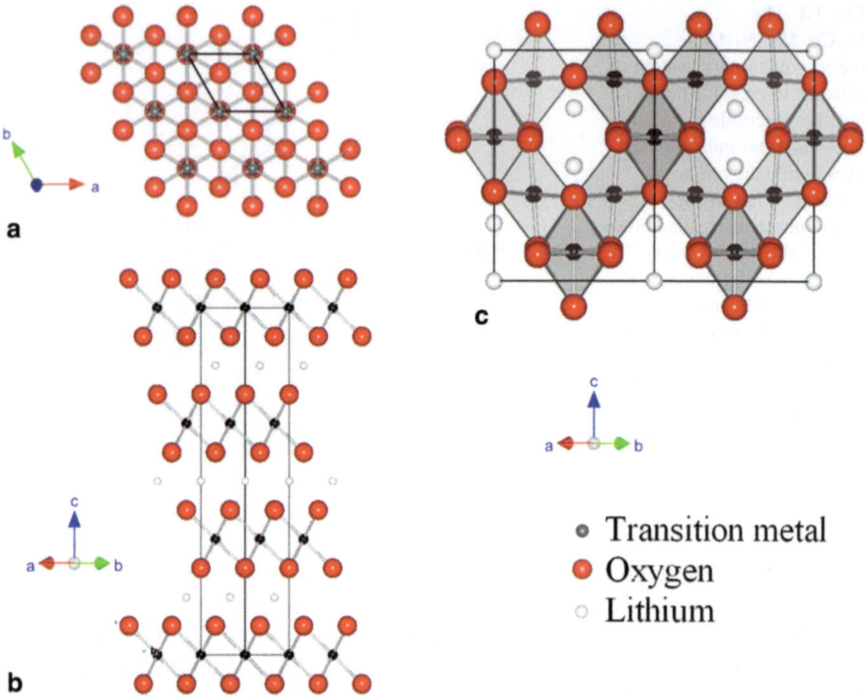

Fig. 1.2 Structure of a layered lithium metal oxide along the 001 projection (**a**) and the 110 projection (**b**). **c** Structure of a cubic spinel along the 110 projection showing the lithium atoms in the tunnels. The bonds between transition metal atoms and oxygen atoms are shown as well as the edges of the unit cell. The octahedra shown in **c** have a transition metal atom at the center and six oxygen atoms at the corners

where the square brackets represent the transition metal layer [1]. This notation will be used throughout this thesis in situations where it is useful to distinguish between the lithium layers and the TM layers. Barring any vacancies or other defects, these materials have a one-to-one oxygen-to-metal ratio. It is important to notice that this idealized structure changes considerably as compositions are altered, e.g., some nickel is often present in the lithium layer and lithium can be found in the TM layer [2].

Charging a Li-ion cell with a positive electrode having a layered oxide structure removes lithium by way of two different mechanisms. The first is deintercalation of lithium wherein the oxidation number of one of the transition metals increases in order to maintain an overall zero oxidation number in the electrode material. The second mechanism is less well understood and takes place around 4.5 V in materials with excess lithium and results in some lithium remaining after all transition metal atoms have been oxidized [2]. This capacity near 4.5 V, referred to here as the "high voltage plateau," is accompanied with a phase transformation [15, 16] and may also involve the production of oxygen gas [17]. The amount of oxygen loss may depend on the stoichiometry of the starting material [15]. Although some electrochemical studies are included in this thesis, the primary focus will be the structures of the starting materials such that an in-depth description of the electrochemical behavior of these materials will not be included in this introduction. It is sufficient to point out that lithium can be removed from the layered materials by way of 2-D transport along the hexagonal planes.

Figure 1.2c shows that the spinel structures have lithium atoms aligned in rows within the cubic lattice. Here, every lithium atom is in a tetrahedral site, surrounded by four nearest neighbor oxygens. In spinel structures, the oxygen content is higher than for layered materials as there are four oxygen atoms per three metal atoms. This higher oxygen content in the spinels will be significant in discussing how the boundaries of the single-phase regions move under various synthesis conditions. Charging of a cell made with a spinel positive electrode involves one dimensional Li-ion transport through the 3-D network of tunnels and proceeds by the deintercalation mechanism described for the layered structures [1].

While no attempt has been made to precisely work out electronic structures of the layered and spinel materials, simple models considering the oxidation state of each metallic atom can be very useful in predicting the possible mechanisms taking place during cycling. To illustrate the importance of the oxidation states of the starting materials, consider $LiMn_2O_4$. The initial average oxidation number of the manganese is 3.5+ such that all the lithium can be removed while the manganese transitions to the 4+ state. However, it is possible to synthesize $Li_{1+x}Mn_{2-x}O_4$ materials with more lithium, up to a maximum of $Li_{4/3}Mn_{5/3}O_4$ [7, 8]. In this material, all manganese is in the 4+ state such that no redox transition is possible and thus no lithium can be removed in the potential window typically used (below 5.0 V vs Li^+/Li). This illustrates how very simple models can be useful. This approach will be used throughout this thesis to discuss the electronic structures of the metals in these complex structures as they undergo changes during electrochemical cycling.

1.3 The Li–Co–Mn–O Face of the Pyramid

Figure 1.3 shows the single-phase structures known in the Li–Co–Mn–O system prior to the current work. To demonstrate how these Gibbs triangles are constructed, consider the sample $LiCo_{2/3}Mn_{4/3}O_4$ that has metallic fractions $Li_{1/3}Co_{2/9}Mn_{4/9}$ and is, therefore, plotted at the point (Li, Co, Mn) = (0.333, 0.222, 0.444). The tick marks on all ternary diagrams are slanted to make the values of the three axes more apparent. Only the metal atoms obey the rules of a Gibbs triangle and the three axes are therefore the metal atomic fractions with Li + Co + Mn = 1 for all points. The labels at the three corners only show the phases present at the corners. Also, since the cobalt content can be calculated as 1 − Li − Mn, all points will be shown as (Li, Mn), such that (0.333, 0.444) refers to $LiCo_{2/3}Mn_{4/3}O_4$, (0.333, 0.666) refers to Li_2MnO_3, and (0.5, 0) refers to $LiCoO_2$. It is worth noting that since the oxygen content varies through the triangle, this pseudo-ternary diagram represents a nonplanar surface

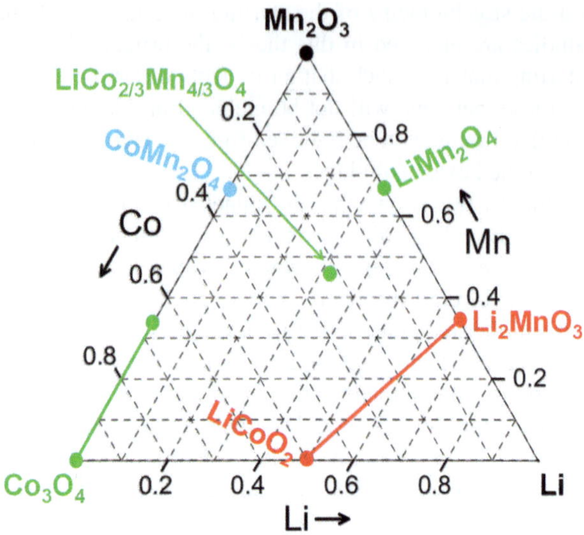

Fig. 1.3 The Li–Co–Mn–O pseudo-ternary system with single-phase materials known prior to the combinatorial studies discussed here. The *red* symbols indicate layered oxides while the *green* and *blue* are cubic and tetragonal spinels, respectively, and the *black circle* represents bixbyite Mn_2O_3. The lithium corner is labeled as Li as no stable solid phases are seen here as lithium is gradually lost during synthesis

in the Li–Mn–Ni–O quaternary system. As such, the oxygen content at any point is not only a function of the position within the triangle but also depends on the synthesis conditions. The atmosphere and temperature profiles used during heating are particularly important.

Figure 1.3 also shows that only a few spinel structures were known in this triangle prior to this research [19–21]. However, the fact that the cobalt corner takes a cubic spinel structure, Co_3O_4, and the placement of the other single-phase spinel materials, suggests that there may be a large spinel region spanning from $LiMn_2O_4$ to Co_3O_4.

The layered structures in the Li–Co–Mn oxide system have been studied extensively along the composition line $Li[Li_{(1-x)/3}Co_xMn_{(2-2x)/3}]O_2$; $0 \leq x \leq 1$, which joins $LiCoO_2$ to Li_2MnO_3 [18, 22, 23]. Figure 1.4 shows X-ray diffraction (XRD) patterns along this composition line reported by Kim et al. [18]. The XRD pattern for $LiCoO_2$ can be attributed to the layered structure described in the previous section where nearly every site on the TM layer is occupied by cobalt. All scattering peaks can be indexed to the R-3m space group, which has a hexagonal lattice with every unit cell being made up of three TM layers and three lithium layers (this stacking is referred to as O3-type and is illustrated in Fig. 1.2b). Figure 1.5 shows two possible arrangements of atoms on TM layers. The first, shown in Fig. 1.5a, assumes random occupation of all sites as would be obtained for $LiCoO_2$. For Li_2MnO_3, however, there are extra peaks seen in the scattering angle range 20–30° which can be attributed to ordering on the TM layer [24]. Figure 1.5b shows such an ordered layer where the red atoms represent lithium while the blue atoms represent manganese. The red atoms all lie on a lattice that is $\sqrt{3}$ times larger in each direction than the hexagonal lattice shown in Fig. 1.5a. As such, this ordered lattice will be referred to as a $\sqrt{3} \times \sqrt{3}$ superlattice. The ordering on the TM layer is typically accompanied by a distortion in the stacking of the layers such that the structure can no longer

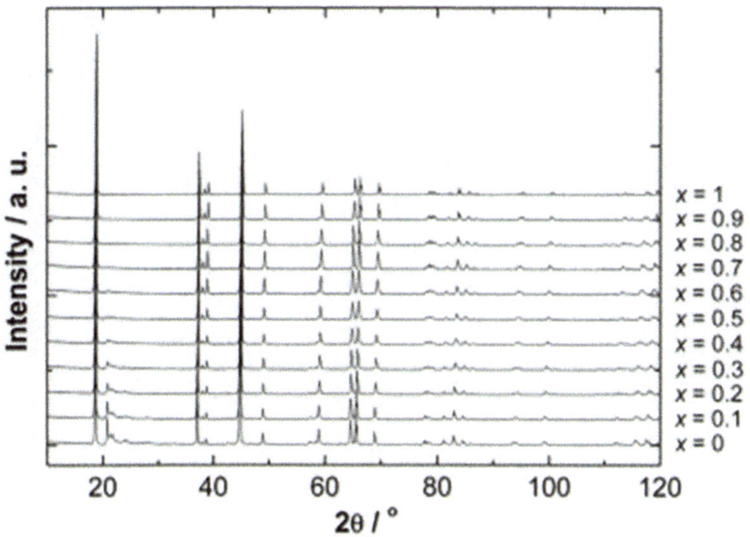

Fig. 1.4 Stack of XRD patterns for various values of x along the line from Li_2MnO_3 ($x = 0$) to $LiCoO_2$ ($x = 1$). *Vertical offsets* are applied for clarity, as is the practice throughout this thesis. (Reprinted from Ref. [18] with permission from Elsevier)

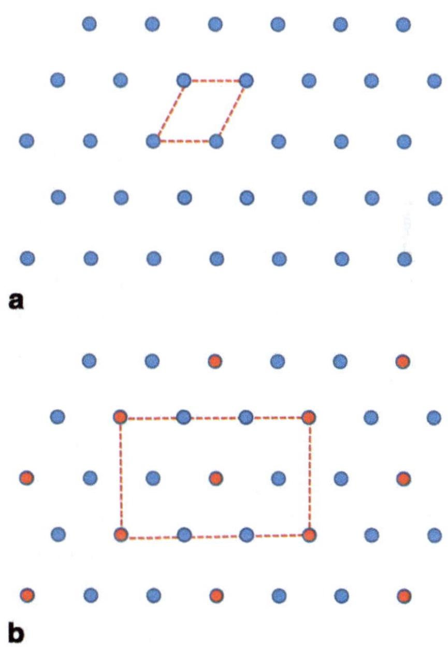

Fig. 1.5 **a** A hexagonal array of atoms as seen in layered Li–Mn–Ni oxides as would be obtained for a disordered transition metal layer where every site is randomly occupied by Ni, Mn, and/or Li. The *red dashed lines* indicate the unit cell. **b** A transition metal layer showing ordering where one third of the sites are *red* (e.g., lithium in Li_2MnO_3) and the other two thirds are *blue* (e.g., manganese in Li_2MnO_3). The *red dashed lines* show the monoclinic unit cell necessary to describe the structure

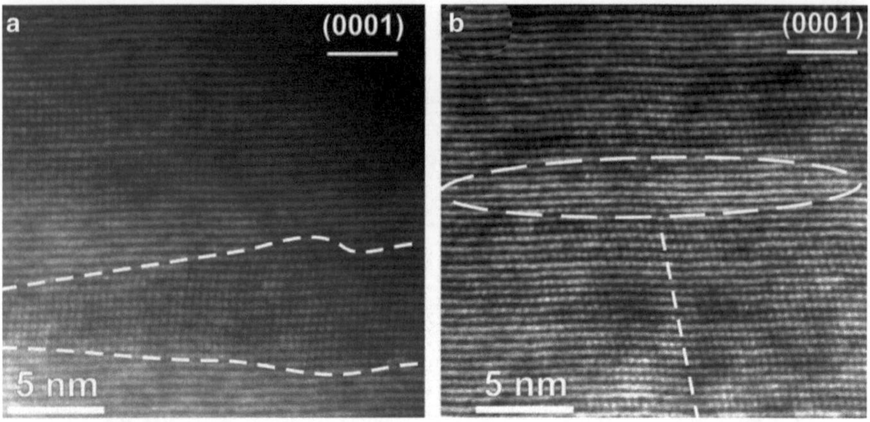

Fig. 1.6 TEM images of $Li_{1.2}Co_{0.4}Mn_{0.4}O_2$ showing the coexistence of layered and monoclinic domains on the nano-scale. (Reprinted from Ref. [22] with permission from Elsevier)

be described as a hexagonal lattice, but instead a monoclinic unit cell is required. Typically, the C2/m space group is used to describe these structures. However, the monoclinic distortions in the lattice for Li_2MnO_3 are relatively small and require long annealing times at high temperature to develop extended long range order [24]. These distortions can best be seen in a peak splitting in the (104) hexagonal peak near 45° in the XRD patterns, which is replaced by the (− 202) and (131) peaks in the monoclinic structure. Some of the XRD patterns obtained in this thesis were refined for both space groups in order to determine where monoclinic distortions became significant.

There is no consensus in the literature as to whether the line from $LiCoO_2$ to Li_2MnO_3 forms a solid solution [18] or if there is phase separation taking place on the nanometer length scale [23]. Kim et al. found that the XRD patterns were consistent with a solid solution with peaks indexing to either a hexagonal space group, R-3m (over the entire composition range), or to the ordering peaks in the C2/m space group, which increased in intensity as the Li_2MnO_3 endpoint was approached. However, careful examination of the published XRD patterns from Kim et al. (Fig. 1.4) shows that the peaks at high scattering angles broaden near the center of the line (near x = 0.4–0.6). This broadening could either be due to a reduction in crystallite size [25] or to the coexistence of two closely related structures. One objective of this project is to determine the actual source of the peak broadening and to develop tools to help distinguish these two situations as this occurs repeatedly in these systems **I**.

In contrast to the solid-solution model, studies by Wen et al. [22] and Bareño et al. [23] show transmission electron microscopy (TEM) and X-ray absorption near edge structure (XANES) data suggesting layered–layered phase separation takes place on the 2–10 nm length scale for a sample made at 900 °C with composition $x = 0.4$. This sample lies in the region where peak broadening can be seen in the XRD data published by Kim et al. [18]. Figure 1.6 shows TEM data demonstrating that two

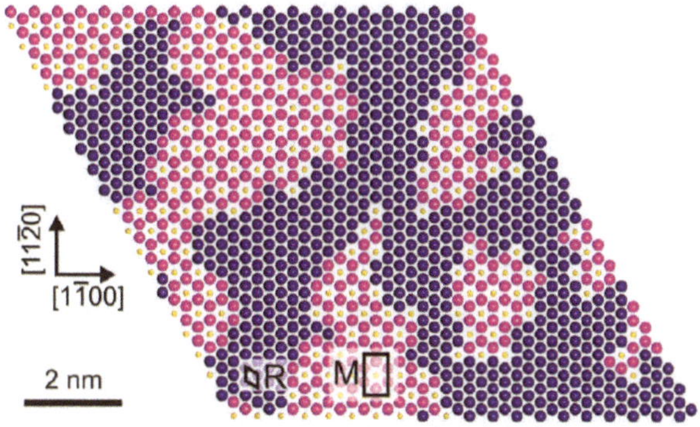

Fig. 1.7 A schematic of layered–layered phase separation with domains of Li_2MnO_3 and $LiCoO_2$. Lithium atoms are *yellow*, while manganese atoms are *purple* and cobalt atoms are *blue*. (Reprinted from Ref. [23] with permission from the American Chemical Society)

different domains, separated by dashed lines, exist. The domains made up of rows of pairs of atoms are attributed to Li_2MnO_3 where only two-third of the sites occupied by manganese appear in the image. Figure 1.7 shows the outcome of a simple model used to illustrate the phase separation on the TM layer. The model was made by Bareño et al. [23] assuming the composite material was made up of domains of pure $LiCoO_2$ (Co on TM layer) and Li_2MnO_3 ($Mn_{2/3}Li_{1/3}$ on the TM layer). For both of these phases, the lithium layer is filled solely with lithium atoms such that the phase separation does not affect the lithium layer. However, on the TM layer, the result yields nano-scale domains of each phase. An important question to be answered is whether or not the presence of these nano-scale domains can be detected in the XRD patterns. This depends greatly on the difference in lattice parameters between the two phases (if they are sufficiently different, the phase separation should at the very least result in severe peak broadening). Since TEM data cannot be used to determine the compositions of each phase present, it will be important to determine whether or not phase separation does occur over the entire $LiCoO_2$–Li_2MnO_3 composition line. Chapter 4 of this thesis will deal with this region of the Gibbs trangle and will help determine over which composition ranges the phase separation actually occurs and which conditions give rise to the co-existence **II**.

1.4 The Li–Mn–Ni–O Face of the Pyramid

Figure 1.8a shows the Li–Mn–Ni–O pseudo-ternary phase diagram with single-phase regions that had already been studied extensively before this project. Again, as in the Li–Co–Mn system, points in the Gibbs triangles will be denoted by two coordinates: (Li, Mn). Here, the nickel metal content is $1 - Li - Mn$. It warrants pointing out that computational combinatorial science has already been invaluable in screening for potential electrode materials [26–28]. However, Fig. 1.8b shows that the public materials database [29] is currently limited to the binaries at the outer edges of the triangle and no solid solution regions extending into the triangle have yet been identified by such methods. Therefore, a large amount of work must yet be done

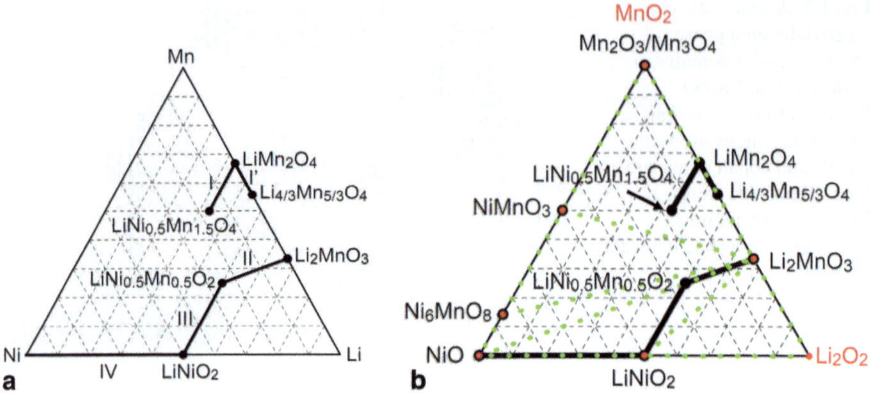

Fig. 1.8 **a** The Li–Mn–Ni oxide pseudo-ternary system where the corners refer to the metals used during sample preparation and oxygen content varies throughout the triangle. The *bold lines* represent lithium-containing single-phase regions. Spinel samples are along *line I:* $LiNi_xMn_{2-x}O_4$ with $0 \leq x \leq 0.5$. Samples along the line labeled I' are spinel at lower temperatures but phase separate at the higher temperatures dealt with here. The *other lines* are rocksalt structures. *Line II* is layered $Li[\,Li_{(1-2x)/3}N_xMn_{(2-x)/3}]O_2$; $0 \leq x \leq 0.5$, *III* is $Li[Ni_{1-x}Mn_x]O_2$; $0 \leq x \leq 0.5$ and *IV* is $Li_xNi_{2-x}O_2$; $0 \leq x \leq 1$. **b** The same diagram with single-phase samples predicted by the public materials database [29] in *green dotted lines* are tie-lines at the edges of three-phase regions

before computations can fill in the system with the accuracy required, particularly in the coexistence regions where identifying what phases coexist is particularly computationally demanding. As such, an experimental approach is used here and this system has been mapped out in its entirety by making samples at over 300 different compositions.

Carey and Dahn [10] successfully used a combinatorial robot to synthesize $LiNi_xMn_{2-x}O_4$ spinel samples (line I in Fig. 1.8) by carbonate coprecipitation of microliter mixtures of lithium, manganese, and nickel nitrates followed by heating to 800 °C in air. It is, therefore, tempting to apply this method to all samples throughout the triangle. However, there is a significant challenge: certain regions show extreme lithium loss during synthesis such that certain sections of the triangle cannot be synthesized even at 700 °C in air. In the literature, this loss is either viewed as Li_2O evaporation [30] or the formation of lithium peroxide vapour [31]. $LiNiO_2$ has a low lithium-binding energy and so it shows some of the highest losses; typically 5 % in bulk samples synthesized by solid-state reactions and heated in air [32, 33]. The small combinatorial samples result in far greater losses due to the larger surface area-to-volume ratios, while the $LiNi_xMn_{2-x}O_4$ spinel samples showed virtually no lithium loss [10] due to the larger lithium-binding energy. Thus, before studying the entire system in Fig. 1.8, the lithium loss in the lithium nickel oxide samples must be minimized. This will be explored in Chap. 3 and will help resolve whether lithium is lost as lithium oxide or peroxide **III**.

Once again, the key structures for Li-ion batteries in the Li–Mn–Ni–O diagram are either spinel or layered. Line I represents spinel samples with composition

$LiNi_xMn_{2-x}O_4$ with $0 \leq x \leq 0.5$ [6, 34], while samples along line I' are spinel at lower temperatures [7, 8] . The other lines are layered structures: line II is the lithium-rich layered line $Li[Li_{(1-2x)/3}Ni_xMn_{(2-x)/3}]O_2$; $0 \leq x \leq 0.5$ [2], III is $Li[Ni_{1-x}Mn_x]O_2$; $0 \leq x \leq 0.5$ [35], and IV is $Li_xNi_{2-x}O_2$; $0 \leq x \leq 1$ [36]. Another composition line in the triangle that has previously been studied in detail is the $Li_xNi_{2-x}O_2$ line from $x = 0$ to 1 (from the Ni corner to $LiNiO_2$). For $x < 0.62$, samples are cubic rocksalt where the metal atoms occupy a face-centered cubic lattice and every metal site is randomly occupied. Though these materials are not interesting as positive electrodes they warrant study because they often appear as contaminants in the synthesis of either spinel materials or layered–spinel composites as discussed below. Above $x = 0.62$, the structures are hexagonal with the lithium and nickel preferentially ordering on alternating layers as illustrated in Fig. 1.2a [36]. The transition at $x = 0.62$ can be identified by fitting the XRD patterns as hexagonal and plotting the c/a lattice parameter ratio versus x. Extrapolating to the point where the ratio reaches $\sqrt{24}$, the expected value for a cubic structure [36] gives the position of the phase transition. This method will be used in Chap. 5.

1.4.1 Li–Mn–Ni–O Spinel Solid Solutions

The spinel solid solution (line I in Fig. 1.8) has been thoroughly studied. The nickel-rich endpoint, $LiNi_{0.5}Mn_{1.5}O_4$, is single-phase when prepared in oxygen but shows a contaminant rocksalt phase when heated in air [34]. More recently, Ma et al. attributed the contaminant to Ni_6MnO_8 [37] while Cabana obtained a cubic lattice parameter of 4.15 Å leaving the stoichiometry of the contaminant phase undetermined [38]. This confusion about the actual composition of this contaminant will be resolved here by mapping out the entire spinel–rocksalt coexistence region carefully **IV**.

It is also important to be mindful that $LiNi_{0.5}Mn_{1.5}O_4$ spinel can sustain oxygen vacancies at high temperature and that oxygen returns into the sample during slow cooling [39, 40]. Compositions of $LiNi_{0.5}Mn_{1.5}O_{4-\delta}$ with $\delta = 0.1$ were obtained at temperatures at or above 750 °C [41]. However, it is hard to distinguish this from phase separation with the formation of the rocksalt phase since both occur simultaneously in air and both result in an increase in the lattice parameter as well as a mass decrease due to oxygen loss. The observations in Chap. 5 comparing quenched and slower cooled samples over wide composition ranges helps distinguish these two forms of oxygen loss **V**.

1.4.2 Li–Mn–Ni–O Layered Solid Solutions

As previously mentioned, the lithium-rich layered oxide structures have alternating lithium and transition metal hexagonal layers, with some excess lithium on the TM layers. Figure 1.9 and Ref. [2] show that when synthesized in air these structures

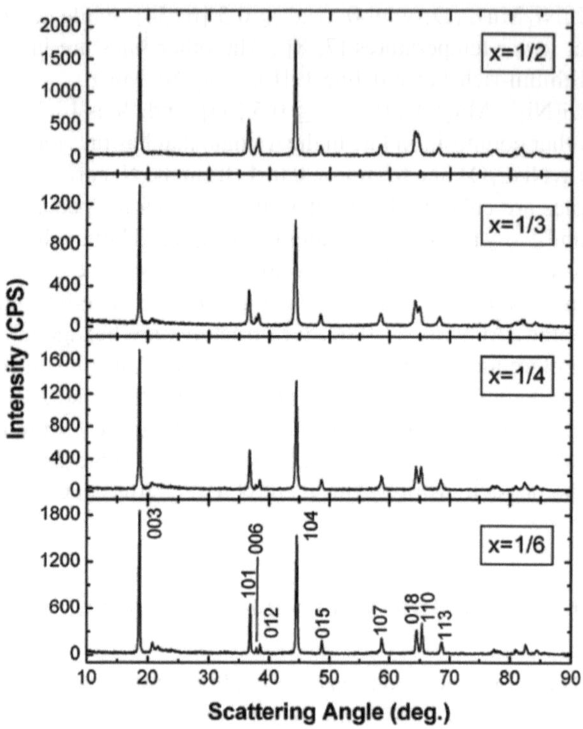

Fig. 1.9 XRD patterns for samples along the composition line from Li_2MnO_3 ($x = 0$) and $LiNi_{0.5}Mn_{0.5}O_4$ ($x = 0.5$). Peaks are indexed according to the R-3m space group for hexagonal structures. (Reprinted from Ref. [45] with permission from the American Chemical Society)

form a solid solution along the whole composition line at or above 800 °C according to the XRD patterns [2]. Once again, although the structures in the layered region are all O3-type, they are not all described by the same space group. For example, Li_2MnO_3 takes a monoclinic structure best described by the C2/m structure while $LiNi_{0.5}Mn_{0.5}O_2$ takes an R-3m hexagonal structure with random occupation on the TM layer [42]. One important point of conflict in the literature is whether or not the lithium-rich line forms a solid solution. Some argue that the superlattice peaks seen in the XRD patterns of samples where the lithium content on the TM layer is not one-third is an indication of local phase separation into Li_2MnO_3 and $LiNi_{0.5}Mn_{0.5}O_2$ domains on the nanometer length scale [43, 44]. Lu et al., however, suggested that the weak ordering peaks can be attributed to ordering of lithium and manganese on the transition metal layers with nickel randomly occupying the remaining sites [45].

Figure 1.10 shows how Lei et al. [44] used TEM data to support the claim that this phase separation occurs in $Li_{1.2}Ni_{0.2}Mn_{0.6}O_2$ while Jarvis et al. [46] presented TEM data supporting that this same structure is made up of a single phase. Figure 1.11 shows TEM images of single-phase Li_2MnO_3 with domains of perfect O3 stacking with the ordering expected for this material. However, there are significant concentrations of stacking faults which seem to occur on roughly the same scale as the new phase identified by Lei et al. in Fig. 1.10b. Clearly, TEM studies of these materials remain inconclusive as to whether or not short range phase separation occurs.

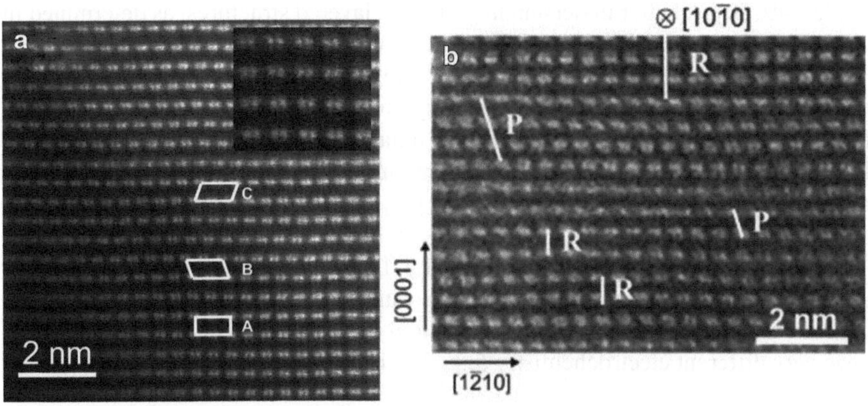

Fig. 1.10 TEM images of $Li_{1.2}Ni_{0.2}Mn_{0.6}O_2$. (**a** Reprinted from Jarvis et al., Ref. [46], with permission from the American Chemical Society. **b** Reprinted from Lei et al., Ref. [44], with permission from Elsevier)

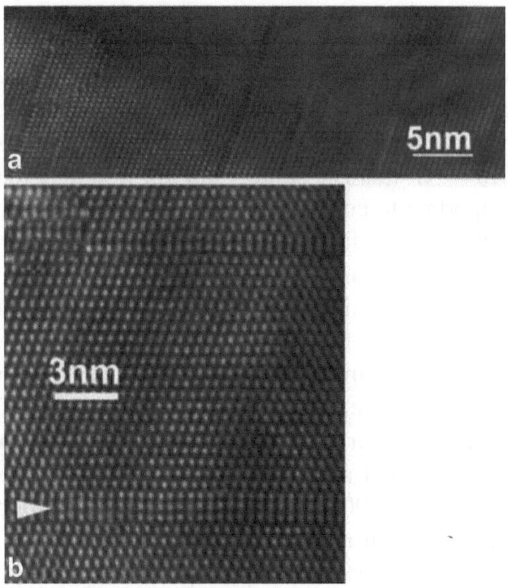

Fig. 1.11 TEM images of single-phase Li_2MnO_3 showing ordered domains with stacking faults, one of which is indicated by the *white arrow*. (Reprinted from Ref. [47] with permission from Springer)

The phase diagrams obtained here offer significant insight into the compositions at which layered–layered phase separation might occur and under which conditions. This work, therefore, helps resolve this issue **VI**.

Recent work, again on the lithium-rich layered oxide $Li_{1.2}Ni_{0.2}Mn_{0.6}O_2$, has suggested that as much as 25 % of the manganese is in the 3 + oxidation state based primarily on the fact that the magnetic moment was lower than that expected if all manganese is in the 4 + state [48]. This result is unexpected since research shows that manganese is typically synthesized in the 4 + state in the lithium-rich layered oxides [49, 50]. A better understanding of these layered structures, as determined in Chap. 8, will help explain these discrepancies **VII**.

The final source of confusion in the literature relating to the lithium-rich structures involves the use of excess lithium during synthesis. As discussed here, the samples lose some lithium during synthesis and so it has become common practice to make lithium-rich layered oxides with a small amount of excess lithium (e.g. 5 %) and to assume the excess burns off during heating [51, 52]. It will be shown here that this is not always the case and some of this excess lithium can be tolerated in the structures such that they may lie below line II in Fig. 1.8. Lattice parameter contour plots prove invaluable in determining the actual compositions of such published samples and this will help explain why materials seemingly synthesized at the same composition have very different electrochemistry (e.g., $LiNi_{0.5}Mn_{0.5}O_2$ [2, 52, 53]) **VIII**.

1.4.3 The Coexistence Region Between the Spinel and Layered Structures

There are a few particularly strange results in the literature for samples made at the composition of $LiNi_{0.5}Mn_{0.5}O_2$. The first is that the XRD peaks of $LiNi_{0.5}Mn_{0.5}O_2$ are very broad when synthesized in oxygen at 800 °C. This broadening of the peaks was interpreted by Jo et al. [54] as being due to small crystallite sizes of about 14 nm. However, this is a small value for a sample made at these temperatures and so it is important to consider the possibility that this broadening is due to multiple phases with slightly different lattice parameters. By looking at samples over wide composition ranges and various synthesis conditions it is possible to distinguish between these two possibilities and so one objective of this thesis is to better understand the observations of Jo et al. [54] **IX**. Figure 1.12 shows an even more peculiar observation made in a sample of $LiNi_{0.5}Mn_{0.5}O_2$ by Hinuma et al. [55]. This sample, made by ion exchange and heating to 1000 °C (pristine), showed a strange phase transformation when annealed at 600 °C. The extra peaks found after annealing were not identified and the sample returned to its single-phase layered structure when annealed at 1000 °C and quenched back to room temperature. This dramatic change between high and medium temperature behaviour was seen repeatedly during the current study when comparing quenched samples to those obtained by slow cooling. Thus, another objective of this study is to explain the results seen by Hinuma et al. [55] **X**.

Finally, another significant motivation for this research is the recent interest in composite electrodes combining spinel and layered structures in the Li–Mn–Ni–O system [13, 14, 56, 57]. In such studies, the samples are assumed to be made up of phases along lines I and II in Fig. 1.8, with the occasional presence of rocksalt contamination (this turns out to be the same rocksalt contaminant as seen in the spinel samples and will be identified in Chap. 5). Figure 1.13 shows the XRD patterns obtained by Cabana et al. [13]. The results clearly show spinel and layered coexistence. However, without clearly knowing the boundaries of the single-phase regions, nor the

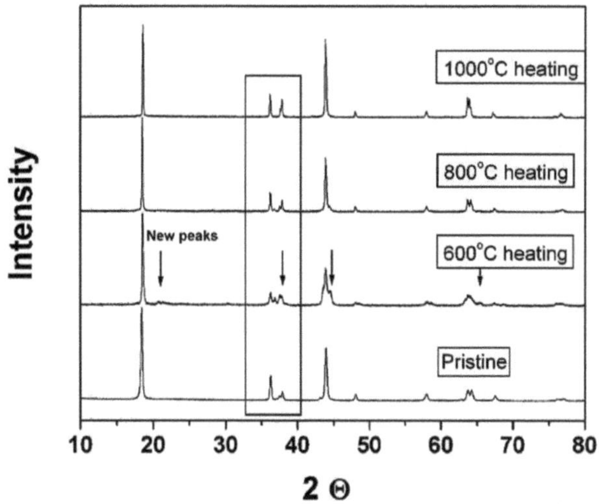

Fig. 1.12 XRD patterns. Samples of $LiNi_{0.5}Mn_{0.5}O_2$ were quenched from high temperature before being annealed at the temperatures indicated. (Reprinted from Ref. [55] with permission from the American Chemical Society)

direction of tie-lines (coexistence regions where all samples along the line are made up of the two phases at the endpoints), it is impossible to determine the composition of each phase and interpreting electrochemical data becomes extremely difficult.

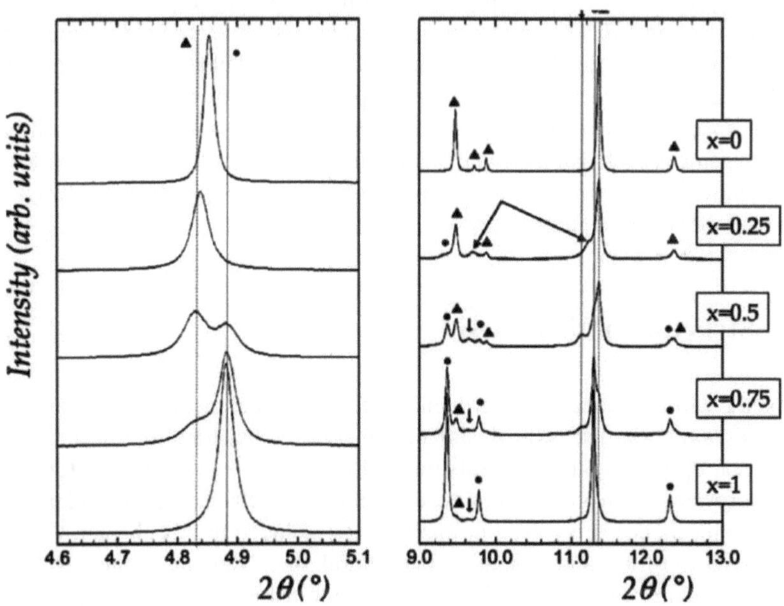

Fig. 1.13 XRD patterns obtained at a synchrotron. Samples lie along the composition line from $Li_{1.2}Ni_{0.2}Mn_{0.6}O_2$ ($x = 0$) to $LiNi_{0.5}Mn_{1.5}O_4$ ($x = 1$). The triangles indicate peaks that index to a layered material, while the circles represent peaks from a spinel material and the arrows indicate a rocksalt contaminant. (Reprinted from Ref. [13] with permission from the Electrochemical Society)

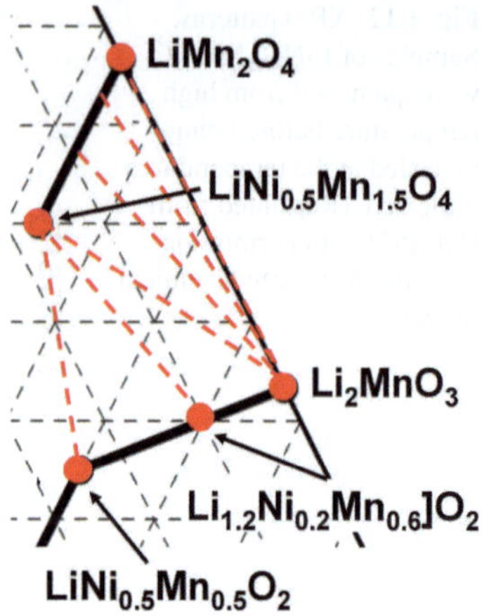

Fig. 1.14 A partial Li–Mn–Ni–O diagram showing the spinel and layered phases known prior to this work (*solid lines*). The *red dashed lines* indicate tie-lines one might expect assuming a two-phase region exists. These tie-lines are refined considerably during this study and are used here only for illustration

Figure 1.14 shows a partial phase diagram with a few proposed tie-lines. This diagram does not hold up well, it is included here for illustrative purposes only. The red dashed lines indicate tie-lines. The line joining $Li_{1.2}Ni_{0.2}Mn_{0.6}O_2$ to $LiNi_{0.5}Mn_{1.5}O_4$ is the tie-line assumed to exist by Belharouak et al. and other groups [13, 14, 56, 57]. However, this tie-line is not accurate as the layered end of the tie-line lies above line II in Fig. 1.8 as will be demonstrated in Chap. 6. Thus, another objective here is to help clarify what phases are actually present in this particularly complex section of the phase diagram **XI**. The current project will precisely determine the nature of the whole coexistence region between the layered and spinel phases which should be of benefit to any researcher studying layered–spinel composite electrodes in this composition space. It is also important to keep in mind that all samples are in equilibrium with oxygen gas such that a sample containing N solid phases is in fact made up of $N+1$ phases. Throughout this thesis, the number of phases present will always refer to the number of solid phases only.

1.5 Structure of this Thesis

Chapter 2 will introduce all experimental and theoretical methods used throughout this thesis, as well as discuss analysis techniques developed for this project. Chapter 3 focuses on optimizing the synthesis method used to make the combinatorial samples. This was done in order to minimize lithium loss such that significant knowledge about the mechanisms involved in this loss are included in this chapter. Chapter 4 presents results from a combinatorial structural study looking at the Li–Co–Mn–Ni–O system and emphasis is placed on phase transformations taking place in the layered structures when slow cooled. Chapters 5 and 6 present the large amount of evidence supporting the Li–Mn–Ni–O phase diagrams. Chapter 5 presents the spinel and rocksalt structures as well as the coexistence regions between them. This information is required before considering the complex behaviors of samples containing layered phases which are of highest interest to the Li-ion research community. Nonetheless, the information gained about the spinel and rocksalt structures is of value in its own right as spinel materials are currently being used commercially and rocksalt contaminants are often seen in materials made in this system. Chapter 6 focuses on regions of the triangle where layered phases are present, which includes the single-phase region, three two-phase regions, and two three-phase regions. This complex system also transforms during slow cooling and evidence for this is included. Chapter 7 demonstrates how the Li–Mn–Ni–O phase diagrams generated in the combinatorial studies can easily be adapted for bulk samples made in air. Chapter 8 provides a detailed study of a new class of layered structures in the Li–Mn–Ni–O triangle where a significant concentration of metal site vacancies are found. Chapter 9 examines materials very near the layered boundary in the region near $LiNi_{0.5}Mn_{0.5}O_2$. Phase separation into layered–layered nano-composites is shown to result in poor electrochemistry. Finally, Chap. 10 summarizes the work presented here, addresses the points of confusion outlined throughout this introduction, and makes suggestions as to future work to be done based on the results from this thesis.

Chapter 2
Experimental and Theoretical Considerations

2.1 Sample Preparation

2.1.1 Synthesis of Combinatorial Samples

The method used to synthesize hundreds of samples across the pseudo-ternary systems was to make combinatorial arrays of milligram-scale oxides. Throughout this section, the method used to make samples in the Li–Mn–Ni–O system will be described. The only variation needed to make samples in the Li–Co–Mn–O system is to replace the nickel starting solution with a cobalt solution (both were nitrates in these studies). This method, developed by Carey and Dahn [10], was closely based on that typically used for large scale samples made in a tank reactor. Carey [10] mixed a total of 10 μL of roughly 2 M lithium nitrate (Aldrich, 98 %), manganese nitrate (Sigma-Aldrich, 97 %) and nickel nitrate (Sigma-Aldrich, 97 %) using a Cartesian Pixsys solution-processing robot shown in Fig. 2.1. Figure 2.2 (a) shows how these solutions were dispensed onto an alumina plate (Pi-Kem, 96 %) coated with stearic acid (Aldrich, 96 %) which served to bead the solutions. Carey then added ammonium bicarbonate (Alfa Aesar, 98 %) in excess to cause co-precipitation of Li, Mn and Ni carbonates. After drying at 55 °C, the sample was made up of the mixed carbonate and any other products of the reaction (in Chap. 3, this will be shown to be primarily ammonium carbonate). Carey then heated the samples to 800 °C for 3 h in air to form the oxides. Silicon (100) wafers were then covered in a tacky mixture of Trilene-65 (a polymer mixture made by Lion Copolymer) and cyclohexane. The wafer was placed over the alumina plates and flipped in order to transfer the samples onto the silicon wafer. The final products were shown to be the expected spinel oxides by XRD.

By contrast, for bulk samples, a precursor is made over the course of several hours by mixing manganese and nickel solutions in a tank reactor in the presence of a precipitator such as ammonium bicarbonate or ammonium hydroxide. The results of this co-precipitation reaction, after rinsing, is a mixed manganese–nickel compound (either hydroxide or carbonate) which is then mixed with a lithium salt (typically Li_2CO_3 or LiOH) and heated to high temperature to make the oxide material. The main differences between the combinatorial approach and the tank reactor method

E. McCalla, *Consequences of Combinatorial Studies of Positive Electrodes for Li-ion Batteries*, Springer Theses, DOI 10.1007/978-3-319-05849-8_2,

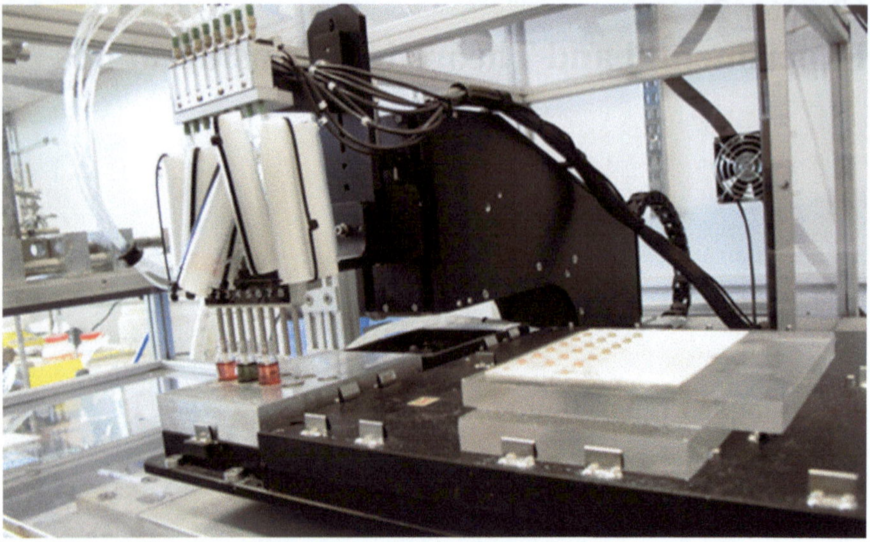

Fig. 2.1 The PixSis solution-dispensing robot used to make all combinatorial samples in this thesis. The three ceramic tips are pulling solutions from the vials in preparation for a dispense onto the partially filled alumina plate

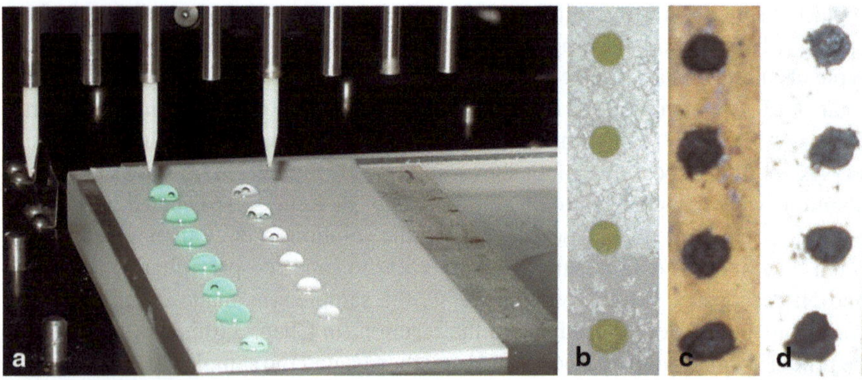

Fig. 2.2 **a** The PixSis solution-dispensing robot making samples with lithium and nickel. Spinel samples at various stages of synthesis. **b** After drying overnight. **c** After heating to 300 °C. **d** After heating to 800 °C in oxygen

are that in the combinatorial approach, all three metals are mixed at the same time, the co-precipitation reaction times are short, there is stearic acid present and the samples are not rinsed after the co-precipitation. The main consequence of the short reaction times is that particles are smaller than those seen typically for bulk samples, especially with the short 3 h heating times. This can be seen by examining the crystallite size in Chap. 3 and in the SEM images in Chap. 6. The fact that particles are small may allow phase separation to occur more rapidly in these samples, thereby confirming the changes seen in the combinatorial samples also occur in bulk samples is critical and is dealt with in Chap. 7. The main consequence of having stearic acid present is that there are contaminants in the samples, though some contaminants are also due to not rinsing the samples after the co-precipitation reaction, and all were found to disappear during synthesis. Figure 2.2 (b)–(d) shows the samples at various stages of heating. After drying at 55 °C, the stearic acid can be seen on the substrate, while at 300 °C, decomposition of the stearic acid begins. The fact that the substrate appears clean after heating to 800 °C shows that decomposition of the stearic acid does reach completion. There is no evidence that the contaminants during heating affect the final products, nor is there any evidence of there being consequences to mixing all three metals in one step, but again it is important to confirm that the tank reactor method gives comparable results to those obtained for the combinatorial samples, and so bulk and combinatorial samples will be compared throughout this thesis.

Chapter 3 shows the results of testing a number of variables in order to optimize the combinatorial synthesis method to minimize lithium loss. These variables were: the substrate used (alumina, alumina coated with $LiAlO_2$ and magnesia), the precipitator used (ammonium bicarbonate or ammonium hydroxide), the atmosphere used during heating (air or pure oxygen), the amount of solutions dispensed (10 or 20 μL) and the temperature at which the samples are heated (200, 300, 400 ... 1000 °C). The sources of all chemicals were the same as those listed above, and ammonium hydroxide was obtained from Fisher while magnesia plates were obtained from Ceramatec. All substrates were first coated with stearic acid. The best combination obtained in that study was used throughout the rest of the thesis and proceeds as follows. Each combinatorial sample, with a mass of approximately 2 mg, was made by dispensing a total of 20 μL of 1.78 M solutions with the solution-processing robot. The concentrations were measured to within 2 % using atomic absorption as described later in this chapter. The three solutions were lithium, manganese and nickel nitrates, and the amounts of each were varied in order to map out the Gibbs triangle. The substrate used during heating was alumina. After dispensing the nitrates, 23 μL of 2 M ammonium bicarbonate was added, thereby ensuring that it was in excess for all samples. The samples were then dried overnight at 55 °C before being heated for 3 h at 800 °C. Some samples were heated in air in a box furnace while others were heated in a tube furnace under a flow of at least 30 mL/min of oxygen. Four cooling rates were used. The first involved turning off the power to the furnace which will be referred to as regular cooling. At high temperatures, this resulted in an approximate cooling rate of 8 °C/min and an overall rate of about 5 °C/min. This cooling method is comparable to that used in the making of commercial electrodes. The second cooling method was to quench the samples by transferring the alumina plate from the furnace onto a steel slab as quickly as possible. A Mastercraft infrared temperature sensor was used to determine that the quenched samples reached 100 °C within 1 min after removal from the furnace. This corresponds to a cooling rate of roughly 10 °C/sec. A second quenching method was used wherein samples were transferred into liquid nitrogen. This cooling occurs at the order of a few seconds such that a cooling rate of about 100 °C/sec was obtained. The final cooling rate, used occasionally, was to cool the samples at a controlled rate of 1 °C/min and will be referred to as slow cooling.

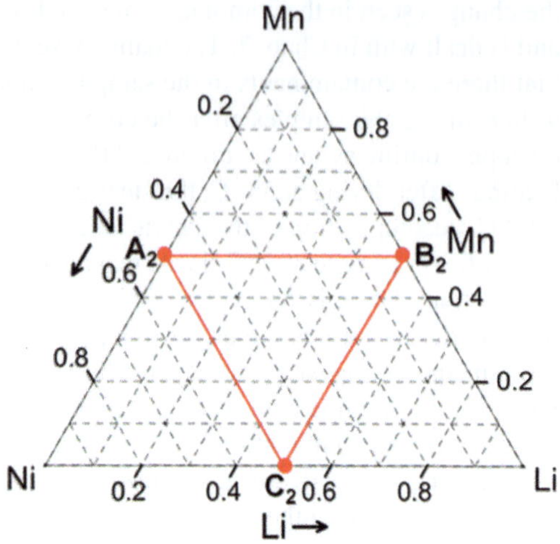

Fig. 2.3 The Li–Mn–Ni oxide triangle illustrating how samples are synthesized over entire composition range in a ternary system

Figure 2.3 illustrates the methodology involved in making samples throughout the Li–Ni–Mn Gibbs triangle. Samples were made at every intersection point of the dashed lines, which resulted in 66 different compositions within the triangle. This was accomplished by making three plates, each holding a 6×6 array of samples. The plates were first cut in half in order to fit into the tube furnace. The three arrays corresponded to samples within each of the three parallelograms shown in Fig. 2.3: $NiA_2B_2C_2$, $LiC_2A_2B_2$ and $MnB_2C_2A_2$. This gave 21 compositions that were repeated on each of the three plates which served as a reproducibility check. Thus, this method results in a total of 108 samples made at 66 compositions. For each system, this mapping was done first. Then, as required, arrays were made over narrower ranges in order to zoom in on compositions of interest. In all cases, there were some duplicate samples to confirm reproducibility (usually duplicating a few samples from the original 66).

2.1.2 *Synthesis of Bulk Samples*

Three different synthesis methods were used to make bulk samples. The simplest was to use a solid-state reaction such as that used in Chap. 3 to make $LiNiO_2$. This involved grinding $NiO_{(s)}$ (Nova Met, 98 %) with $LiOH_{(s)}$ (Sigma-Aldrich, 98 %) and heating to high temperature in air. The second bulk synthesis method mimicked the combinatorial method on a larger scale. This involved mL-scale (usually a total of about 20–40 mL) mixtures of lithium, manganese and nickel nitrates (from the same sources as in the combinatorial synthesis). Excess ammonium bicarbonate was then added to the mixtures in a beaker and stirred for a few minutes. After drying overnight at 55 °C, the samples were heated to 400 °C for about 30 min to drive-off the ammonium carbonate and any other products of the reaction. The samples were then ground and heated for 5 h in air or a flow of oxygen, and either quenched, regular cooled or slow cooled. The resulting samples had masses of approximately 1–2 g. This method will be referred to as the "one-pot" synthesis and was used to determine the impact of the high surface area to volume ratio in the combinatorial samples.

To ensure that the results obtained both by the combinatorial method and the one-pot synthesis are significant for other synthesis routes, bulk samples were also made using a tank reactor as described in Ref. [58]. These samples were made by Aaron Rowe and his contributions to this project will be pointed out explicitly throughout this thesis. A continuously stirred tank reactor was used to make mixed manganese–nickel hydroxide precursors. The metal precursor solutions were made using $NiSO_4 \cdot 6\,H_2O$ and $MnSO_4 \cdot H_2O$ (both 99 % from Aldrich Chemical Co) in such a way that a variety of nickel to manganese ratios were obtained. The tank reactor was used to mix NH_4OH with the metal sulfate solutions at a constant pH of 10.3 throughout the 10 h reaction under a flow of nitrogen. The precursor was then rinsed with distilled water and dried at 140 °C before being mixed with varying amounts of Li_2CO_3 to yield the desired stoichiometry. These mixtures were heated to 900 °C for 10 h in a box furnace and then either quenched to room temperature between two copper plates or cooled at a slower controlled rate of 5 °C/min (which is comparable to the regular cooling rate used for the combinatorial samples and will be referred to as such).

2.2 X-Ray Diffraction

2.2.1 *High Throughput X-Ray Diffraction (XRD) of Combinatorial Samples*

Due to the large number of combinatorial samples, a Bruker D8 Discover X-Ray system was used to characterize the vast majority of the combinatorial samples. All X-ray diffractometers used during this project use Cu–Kα radiation. The Bruker system has a collimated 0.5 mm wide beam, a Göbel mirror and an area detector. Each scan was made up of three frames with a 30 % overlap and each frame is obtained by counting for 300 s. Figure 2.4 shows typical results after stitching the frames together as well as the outcome of integrating along the arcs. This gave a scattering angle range of 15–70 ° and required approximately 15 min per sample. The Bruker D8 diffractometer is equipped with a sample translation stage that allows motion in the x, y and z directions such that the scattering patterns from samples in the combinatorial arrays were automatically measured in sequence. As described previously, the samples were first transferred onto a silicon (100) single crystal which served as a zero-background holder during the XRD scans as long as the Bragg condition for the Si(400) peak was avoided. Although the peak positions determined using the Bruker diffractometer were extremely accurate, the scattered intensities were affected by the stitching of frames and the integration along arcs. As such, Rietveld refinement generally failed for scans produced in this way. The next section describes the diffractometers used when Rietveld refinement was essential.

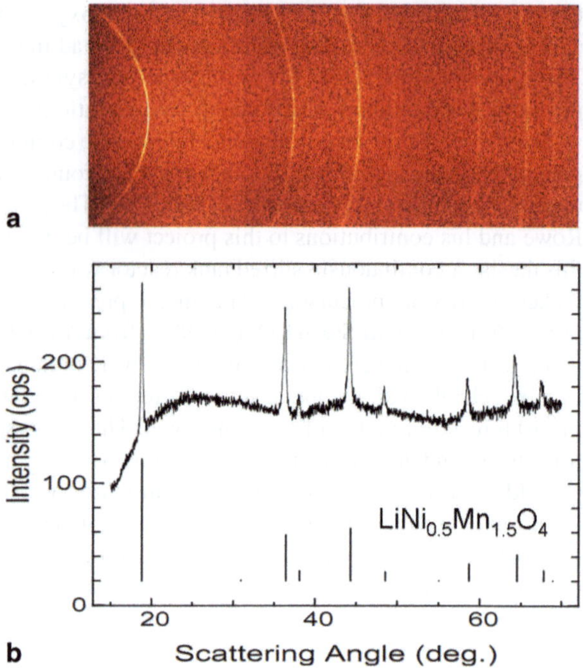

Fig. 2.4 **a** The X-Ray diffraction (XRD) image obtained after stitching three frames together for a spinel sample. **b** The result of integrating along the arcs while the vertical lines correspond to $LiNi_{0.5}Mn_{1.5}O_4$ spinel (JCPDS #80-2162)

2.2.2 *X-Ray Diffraction of Bulk Samples*

The X-ray scattering patterns from bulk samples as well as from a few combinatorial samples were collected using either a JD-2000 diffractometer or a Siemens D-5000 diffractometer. Both are equipped with a Cu-target X-ray tube and a diffracted beam monochromator, though the D-5000 also has a Soller slit to minimize out-of-plane scattering. These scans could be fitted accurately using Rietveld refinement to obtain values for lattice parameters as well as site occupations. Rietveld refinement was performed on single-phase samples only using the software Rietica. The information required was the space group and site occupations, though the occupations were often left as variables in the refinements. The function used for the peak shape was pseudo-Voigt (described in the next section). The primary outputs of the refinements were lattice parameters, site occupations, fitted peak widths and fit quality parameters such as the Bragg factor. In order to account for the average vibrations of the atoms, an overall thermal parameter was allowed to be refined.

2.3 Fitting of Combinatorial X-Ray Diffraction Patterns

Several challenges present themselves with respect to the analysis of XRD data obtained from the high-throughput Bruker machine. The peak intensities are distorted such that Rietveld refinement fails. Carey extracted precise lattice parameters by fitting the top of peaks only [10]. However, this is not feasible when there are multiple phases present for which overlapping peaks result in shifts in apparent peak positions and only using peaks that have no overlap would result in poor precision. To overcome this, an in-house software was written in Yorick to fit the entire scans using a nonlinear least squares fitting algorithm [59]. Fitting the entire patterns had the added benefit of extracting peak width information that was used to calculate crystallite sizes and strains.

Using a degree six polynomial to describe the background was found to converge very slowly. This was resolved by using a cubic function and two broad asymmetric Gaussians centered near 20 and 45 ° to describe the background. This procedure was found to work well for all samples, including spinel samples where manganese fluorescence gives rise to a complicated background when the three frames are stitched together as shown in Fig. 2.4 (b).

In the program, each experimental peak was described by the convolution of the sample scattering with the machine broadening as outlined by Warren [60]:

$$I_k(2\theta) = \int F_k(2\theta - z)M(z)dz \tag{2.1}$$

where 2θ is the scattering angle, F_k is the sample scattering due to the k-th peak, and M is the machine broadening normalized to have an area of unity. The integration over z, the scattering angle of the machine broadening, was done numerically. The sample scattering was described with a pseudo-Voigt function:

$$F_k(2\theta) = I_k^{\circ}[\eta C(2\theta, 2\theta_k, H_k) + (1 - \eta)G(2\theta, 2\theta_k, H_k)] \tag{2.2}$$

where I_k° is the integrated peak intensity of the k-th peak, H_k is the full width at half maximum (FWHM), $2\theta_k$ is the position of the center of the peak, η is the Lorentzian component and is kept constant for all scattering angles, and C and G are the Lorentzian and Gaussian functions, respectively, each normalized to have an area of unity. The peak intensities were fitting parameters, thereby avoiding problems with distortions due to the stitching of frames. The position of each peak was calculated from the h, k and l values given in the Joint Committee on Powder Diffraction Standards (JCPDS) database and from the adjustable lattice parameters. The Kα_1 and Kα_2 peaks were included in a 2:1 ratio in the calculation in order to fit the high angle peaks accurately.

The machine broadening function, M, was determined by measuring the scattering from corundum (NIST standard 1976a), the expected scattering from which is given in JCPDS #46-1212. The scan was fit using pseudo-Voigt functions with $\eta = 0.5$ and letting the FWHM of each peak, $H(\theta)$, be a fitting parameter. Although corundum is not commonly used as a standard for machine broadening, it is sufficient to describe the extensive broadening obtained with the 0.5 mm wide beam. This was determined by measuring the scattering from a silicon wafer which was ground. The results gave peak widths comparable to those obtained with the corundum, however the five peaks showed far more scatter such that corundum proved to be the better standard. Figure 2.5 shows that the resulting peak widths can be described by the Williamson–Hall expression : $H(\theta) \cdot cos\theta = A + Bsin\theta$, where A and B are constants. The slope, B, is negative because the scattering volume decreases for frames with higher incident angles and there is no analyzer monochromator. There is also an evidence of slight plateaus in the Williamson–Hall plot corresponding to the stitching of frames, but this problem is minimized by the 30 % overlap, and the linear fit does describe the data well over the scattering angle range, 15–70 ° used throughout this project.

In order to extract values for crystallite size and micro-strain, the integral breadth method outlined by Klug and Alexander was used [25]. This involved assuming that the size contribution was Lorentzian and describing the strain broadening with a Gaussian function. The integral breadth of the kth peak was obtained by dividing the area of the pseudo-Voigt function by its height, which yields:

$$\beta_k = H_k/\alpha \tag{2.3}$$

where

$$\alpha = 2[\eta/\pi + (1-\eta)\sqrt{(ln2/\pi)}]. \tag{2.4}$$

Utilizing the quadratic approximation [25, 61], and solving for the full width at half maximum gives:

$$H_k = \frac{\alpha}{2}\left[\frac{K\lambda}{Lcos\theta} + \sqrt{\left(\frac{K\lambda}{Lcos\theta}\right)^2 + 64e^2\,tan^2\,\theta}\right] \tag{2.5}$$

where L is the average crystallite size, the root-mean-square micro-strain is $\sqrt{<e^2>} = e/1.25$, and the usual approximation of $K = 1$ was made. The quadratic approximation is generally considered to give values for size and strain that are close to those obtained by Fourier methods without requiring extensive computations [61]. The micro-strain includes all contributions due to nonuniform lattice distortions, dislocations, stacking faults and local structural defects such as vacancies and interstitial atoms. In the case of $Li_xNi_{2-x}O_2$, fluctuations in lithium content within a grain have been observed in the early phases of synthesis in samples with lower lithium content (x < 0.6) [62]. A nonhomogeneous lithium distribution within grains would result in lattice distortions, thereby contributing to the strain. As a result, the strain parameter is of particular interest during synthesis of the samples and is therefore followed closely in Chap. 3.

In order to estimate the minimum crystallite size that can be resolved with the combinatorial X-ray machine, the broadening of the machine peaks due to crystallites of various sizes was calculated by using the Scherrer equation and assuming the peaks were Gaussian to minimize the broadening. A similar calculation was made for strain, assuming crystallites of infinite size. Figure 2.5 shows the results of these calculations and shows that the Bruker XRD system resolves crystallite sizes precisely up to 100 nm; however, above this point, machine broadening begins to dominate and any size greater than 150 nm cannot be determined. For larger crystallites, it would be necessary to use a smaller beam spot (0.3 mm or even 0.1 mm are commonly used), but this was avoided here, since it would have greatly increased the required count

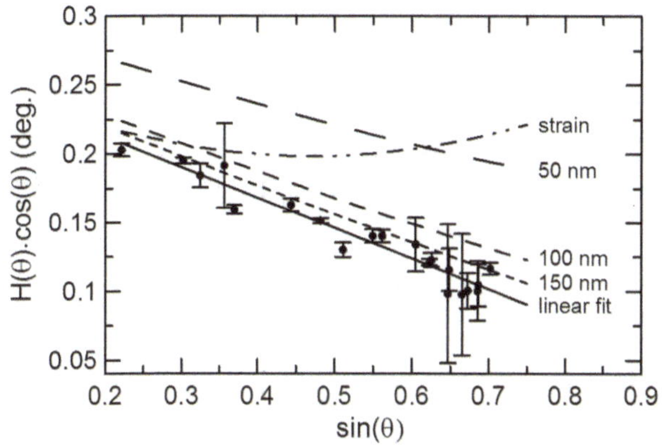

Fig. 2.5 Williamson–Hall plot for corundum obtained with the Bruker XRD machine with a beam spot of 0.5 mm. The *solid line* is a linear fit and is used to define the machine broadening. *Dashed lines* are obtained assuming peaks are Gaussian and using the Scherrer equation to describe the size broadening. The *dotted–dashed line* is obtained similarly, assuming an infinite crystallite with a strain of 1.0x10^{-3}

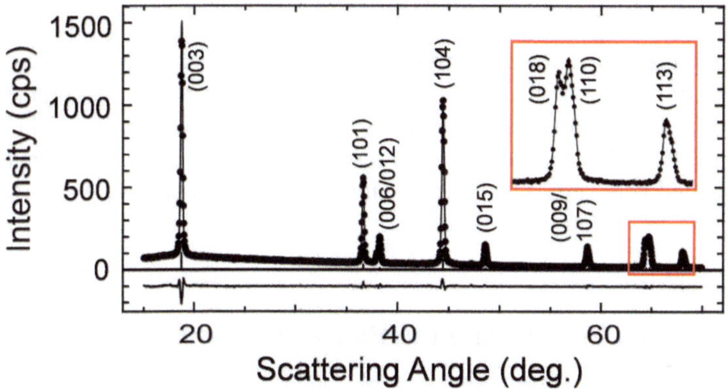

Fig. 2.6 The fit obtained for a combinatorial sample of $LiNiO_2$. This reflects the quality of fit typically obtained for XRD scans of the combinatorial samples taken with the Bruker diffractometer. For clarity, only every third data point is included as *closed circles* while the result of the fit is shown as a *solid line* and the difference plot is shown below

time. Figure 2.5 also shows that a strain value of 1 x 10^{-3} can be readily resolved. Any strain value below 0.1 x 10^{-3} was considered too small to be measured accurately.

Figure 2.6 shows a typical scan obtained for a combinatorial lithium–nickel oxide sample measured with the Bruker machine along with the fit produced by the in-house software. The peaks fitted the experimental data extremely well such that the lattice parameters and the size/strain values extracted from the fitting routine were well constrained. For clarity, all XRD patterns shown throughout the rest of this thesis have the fitted background subtracted from the experimental patterns. This allows for easy comparison of many samples stacked in the same plot.

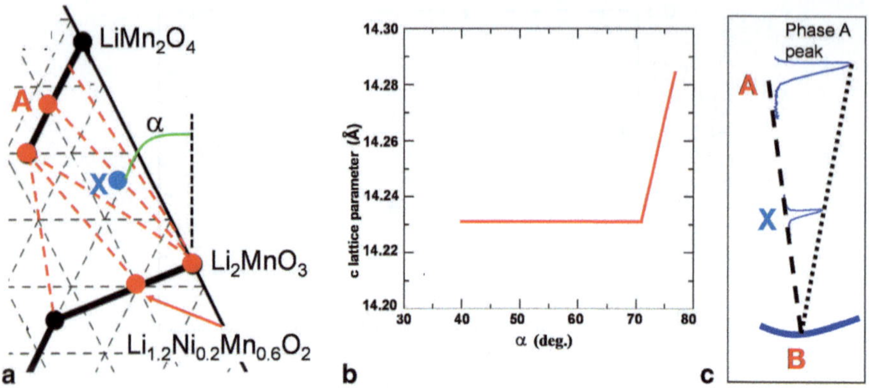

Fig. 2.7 **a** A section of the Li–Ni–Mn–O phase diagram showing previously known single-phase regions. *Red dashed lines* represent proposed tie-lines. **b** An illustration of how lattice parameter values can be used to determine the direction of tie-lines in two-phase regions. **c** A schematic of using the lever rule in a two-phase region to determine the location of a point *B* on the boundary of the single-phase region (e.g. if **a** is correct, point *B* should be Li_2MnO_3)

The program was written to be able to fit multiphase samples. The calculated pattern was therefore a linear combination of the functions representing each of the single phases. Thus, the fitting of multiphase scans generated lattice parameters and integrated peak areas for each phase. The size and/or strain was examined in Chaps. 3 and 4 only. For all other combinatorial results, phase compositions (obtained from the calculated peak areas) and lattice parameters were extracted. As shown in the next section, this information is sufficient to generate precise phase diagrams.

2.4 Methods to Generate the Phase Diagrams

The information extracted from the XRD scans includes: phase types (layered, spinel, rocksalt...), lattice parameters and the peak areas of each peak for all phases present. These are sufficient to determine the boundaries of the single-phase regions and the nature of the co-existence regions. Figure 2.7 (a) shows the proposed tie-lines lying between the spinel and layered regions in the Li–Mn–Ni–O system as discussed in the introduction. This will be used here to illustrate how the XRD patterns could be used to confirm that the proposed diagram was correct. Firstly, all XRD patterns obtained would have to be visually consistent with the phase diagram. This simply means that no signs of a second phase existed in scans of samples in the single-phase region, and no samples lying in the co-existence regions appeared single-phase (as will be shown in Chaps. 5 and 6, this proposed phase diagram fails on both counts). However, visual inspection of XRD scans was insufficient to precisely identify the boundaries of the single-phase region. The next step was to carefully examine the values of the lattice parameters obtained in the co-existence region and to compare them to those found for single-phase samples. In the proposed phase diagram, tie-lines fan out from Li_2MnO_3 up to an angle of about $\alpha = 70°$. This implies that up to this angle, the value of the layered lattice parameters in the two-phase samples should be equal to those of Li_2MnO_3. Figure 2.7 (b) illustrates the expected plot of the c lattice parameter as a function of α based on the proposed tie-lines. Since c = 14.23 Å for Li_2MnO_3 when fitted as a hexagonal structure, as shown in Chap. 6, such a graph would confirm that the tie-lines fan-out from Li_2MnO_3. Plots of this type were extremely useful in demonstrating the directions of tie-lines in the Li–Mn–Ni–O system and were used throughout this thesis. In three-phase regions, there is no variation in the phases present; only the amount of each phase varies. As such, the fitted lattice parameters should remain constant throughout the three-phase region.

In order to work out the boundaries of the single-phase regions, the lever rule was used. Figure 2.7 (c) illustrates the use of the lever rule for sample X lying on a tie-line between phases A and B. Extrapolating to where the integrated peak intensity of the phase A peaks would be zero yields point B. If the proposed phase diagram is correct, point B would coincide with Li_2MnO_3. In practice, the distance from A to B was calculated using the mathematical form of the lever rule which states that the fraction of phase A is equal to d(X,B)/d(A,B). This allows for the calculation of the position of point B, which can then be compared to the position of Li_2MnO_3. This same method was used to determine the positions of the corners of three-phase regions as will be demonstrated in Chap. 6. It should be noted that microabsorption effects were not taken into account in the calculations. This is justified for the combinatorial samples with small particle sizes, as will be demonstrated in Sect. 5.4. Therefore, even without Rietveld refinement, the XRD patterns can be used to identify the boundaries of single-phase regions, the direction of tie-lines in two-phase regions and the extent of three-phase regions. The large number of scans in the single-phase regions also provided lattice parameters over wide composition ranges which were expressed as contour plots in Chaps. 5, 6 and 7. These contour plots should be of considerable use to researchers working in these systems.

2.5 Electrochemical Tests

Coin cells were made from a few bulk samples in order to determine the electrochemical performance of materials of interest. Figure 2.8 shows the parts involved in assembling a standard coin cell. Electrochemical tests were carried out using a standard 2325 coin cell with the positive electrode being cycled against a lithium counter electrode (Chemetall Foote Corp.). The electrolyte used was 1 M $LiPF_6$ in 1:2 ethylene carbonate:diethyl carbonate (Novolyte Corp., now part of BASF). The positive electrodes contained 5 % polyvinylidene-fluoride binder (Kynar 301P, Elf Atochem) and 5 % carbon black (TIMCAL). Two identical cells were typically made. The details of the coin cell design are included in Ref. [63]. The cycling conditions varied from experiment to experiment and will be stated explicitly in the results sections.

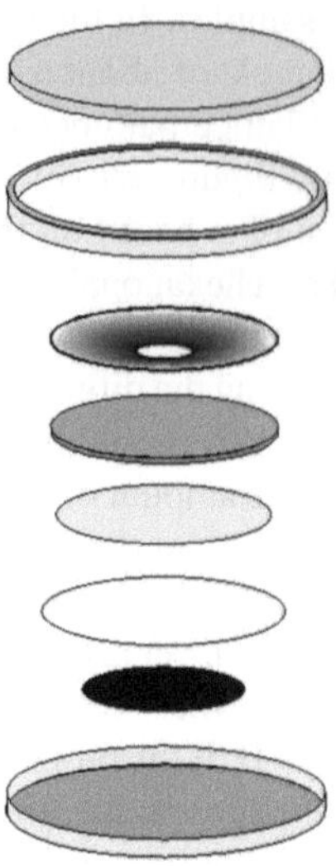

Fig. 2.8 A schematic of the components used in making a typical coin cell for electrochemical testing

2.6 Thermo-Gravimetric Analysis

As a consequence of the high surface area to volume ratio, the combinatorial samples were very sensitive to surface reactions. The mass changes resulting from these reactions were studied using a thermo-gravimetric analyzer (TGA). The TGA was useful in the study of lithium loss during synthesis as will be discussed in Chap. 3. A typical TGA run involved heating approximately 10 mg of sample in a small alumina cup in a TA Instrument, SDT-Q600 TGA under a gas flow of 50 mL/min. These samples were approximately five times larger than those made by the solution-dispensing robot and so the TGA results can be considered to be an ideal limit for the behavior of the combinatorial samples (i.e., the combinatorial samples would lose slightly more lithium that those heated in the TGA). The lithium loss experiments were repeated in flows of argon, air and oxygen in order to identify the role played by the atmosphere.

2.7 Elemental Analysis

With lithium being lost during heating of the combinatorial samples in certain regions of the Li–Mn–Ni Gibbs triangle, the composition of a sample is not necessarily the same before and after heating. As such, either atomic absorption (AA) or inductively coupled plasma optical emission spectroscopy (ICP) have been used as elemental analysis in order to determine the actual compositions of samples. To do this for combinatorial samples, the samples were transferred into 2 mL vials with the use of a transfer plate that clamps the substrate to the vials thereby, preventing mixing of neighboring samples. To each vial, approximately 0.5 mL of a solution of 25 % nitric acid and 75 % hydrochloric acid was added to dissolve the samples. The solutions were then analyzed by Dan Chevalier (Minerals Engineering Centre, Dalhousie University) to give the concentrations of nickel, lithium and manganese, accurate to 5 % for the small samples. The same method was used for bulk samples, though slightly more powder was used to improve the precision of the measurements. For the starting solutions used with the solution-processing robot, AA measurements were repeated three times such that the uncertainty was 2 %.

2.8 Scanning Electron Microscopy

Scanning electron microscopy (SEM) allows the viewing of particles as small as 50 nm in length. As will be discussed in the next chapter, the crystallites made in combinatorial samples heated to 800 °C typically fall in this range such that the SEM was useful in order to get a better idea of whether or not phase separation had occurred. To this end, a few samples were imaged by Ramesh Shunmugasundaram using a Hitachi S-4700 field emission scanning electron microscope with an accelerating voltage of 10 kV and an emission current of 15 μA.

2.9 Redox Titration

As mentioned in the introduction, the oxidation states of the transition metals in the electrode material have a large effect on the electrochemistry. It proved useful to use redox titrations in order to determine the average oxidation state of the nickel and manganese for the materials discussed in Chap. 8. This involved two redox titrations, both using potassium permanganate. The first involved adding the potassium permanganate to the samples (both dissolved in acid) in order to determine the total amount of manganese present. The second titration involved first mixing the dissolved sample with ferrous ammonium sulfate in order to reduce nickel and manganese to the 2+ oxidation state while oxidizing some iron to 3+ state. The subsequent titration with potassium permanganate then reduced iron back to 2+. The two measurements allowed the determination of the average manganese oxidation state (assuming all nickel was in the 2+ state). As such, a result for the manganese oxidation state greater than 4.0 would imply some Ni^{3+} was present. It is important to notice that this method cannot distinguish between a Ni^{2+}–Mn^{4+} pair and a Ni^{3+}–Mn^{3+} pair, since this approach essentially counts the number of oxidation states above 2+. To make the distinction between these two combinations, manganese X-ray absorption spectra were taken as described in the next section. The titrations were performed by Oliver Schilling of Erachem Comilog.

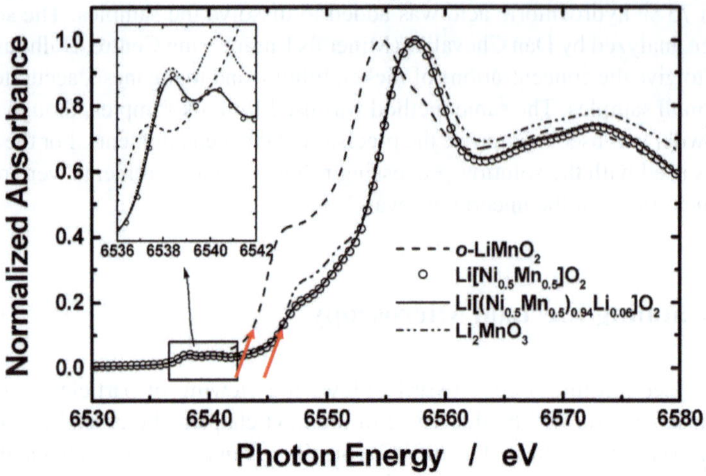

Fig. 2.9 X-ray absorption near-edge structure (XANES) spectra reproduced from Ref. [64] with permission from the American Chemical Society. The *red arrows* mark the rough positions of the main absorption edges

2.10 X-Ray Absorption Spectroscopy

The X-ray absorption spectra of materials are sensitive to both the local and long range electronic structure of the material. Fitting the spectra is rather complex, but information about the oxidation state of each metal can be extracted relatively easily by comparison with known reference materials. Comparing two spectra of materials with similar structures means that the differences seen are primarily due to the electronic configuration of the atom in question.

Figure 2.9 shows the X-ray absorption near-edge structure (XANES) spectra for manganese-containing samples near the Mn K-edge (obtained by exciting 1 s electrons) published by Myung et al. [64]. The position of the main absorption edge, marked for each spectrum with a red arrow, is sensitive to the oxidation state of the manganese atoms. Thus, the two reference materials (Li_2MnO_3 for 4+ and $LiMnO_2$ for 3+) were used by Myung to determine that the manganese atoms in the $LiNi_{0.5}Mn_{0.5}O_4$ sample were in the 4+ state. This simple analysis will be used in Chap. 8 to determine the oxidation state of the manganese atoms. The XANES spectra were collected by Paul Duchesne on the PNC-CAT beamline 20-BM at the Advanced Photon Source of Argonne National Labs. Incident X-rays were made monochromatic using a silicon (111) double-crystal system employing platinum mirrors for harmonic rejection. The samples were all measured in transmission mode at room temperature using ionization chamber detectors.

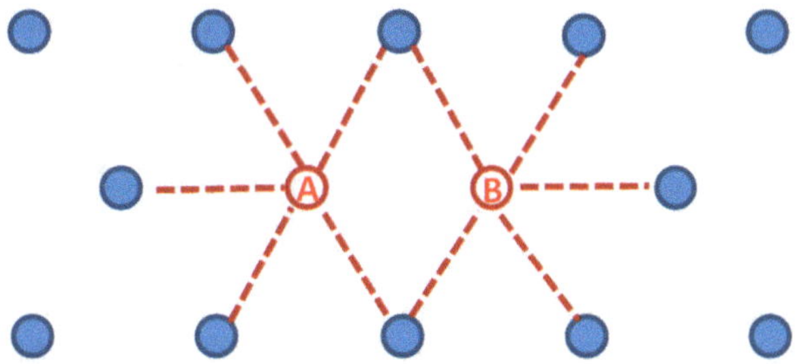

Fig. 2.10 An array of atoms illustrating the interactions used in the Monte Carlo simulations (*red lines*) when calculating the probability of accepting a move wherein atoms A and B are interchanged

2.11 Helium Pycnometry

Some of the materials studied in this thesis showed oxygen non-stoichiometry (either oxygen or metal vacancies). Precise density measurements were used to identify the concentrations of such vacancies. This required knowing the metallic compositions from ICP and the volume of the unit cell from XRD. A Micromeritics AccuPyc II 1340 Helium Pycnometer was used to measure the true volume of the samples. The pycnometer works by pressurizing an empty reference chamber and measuring this pressure before opening a valve between the reference and sample chambers and remeasuring the pressure again. Since the volumes of the empty chambers are known, Boyles' law was used to calculate the volume of the sample. A high precision scale was used to measure the mass of the sample. Multiple measurements with the pycnometer were used in order to calculate a statistical uncertainty. This approach was used in Chaps. 6 and 8.

2.12 Monte Carlo Simulations

As has been discussed, a significant portion of this thesis deals with phase transformations taking place during slow cooling. A Monte Carlo simulation was used to examine the changes that take place on the metal atom layers in the layered oxide materials during slow cooling. This was used on three occasions, in Chaps. 4, 8 and 9. In each case, a 50×50 hexagonal array was made with compositions matching the stoichiometry on the TM layer determined experimentally. The initial array was made assuming random occupation of all sites such that the starting configuration corresponded to infinite temperature where entropy wins out entirely over internal energy. For all simulations, the potential used for the energy of the system was the Coulombic potential for nearest neighbor (NN) interactions only and periodic boundary conditions were used. The effective charge of each atom was assumed to be proportional to its oxidation number. The simulation involved evaluating whether randomly chosen nearest neighbors might exchange position. Figure 2.10 shows the interactions involved in calculating the change in energy between the original configuration and the proposed one obtained by switching the atoms A and B. The effective charge of the i-th atom is $\delta \cdot n_i$, with n_i being the oxidation number of atom i in the original structure and n_i' that of the atom in the structure after the proposed change. The change in energy due to such a move is:

$$\frac{\Delta E}{k_{BT}} = \beta_T \left(\sum_{i,j}^{NN} n_i' n_j' - \sum_{i,j}^{NN} n_i n_j \right) \tag{2.6}$$

where k_B is the Boltzmann constant, T is temperature,$\beta_T = \delta^2/(4\pi\epsilon_o k_B T \cdot a)$, ϵ_o is the permittivity of free space and a is the in-plane lattice parameter. In this model, it was assumed that all nearest neighbors were separated by a distance a. The Metropolis rate equation [65] was used such that the probability of accepting a move is 1 if $\Delta E < 0$ and $e^{-\Delta E/k_B T}$ if $\Delta E > 0$. The simulation was run for increasing values of β_T (0.5, 1, 1.5 ... 5) in order to simulate slow cooling. At each temperature, 10000 Monte Carlo steps were performed (2500 attempted moves constituted a Monte Carlo step).

The third use of this simulation, discussed in Chap. 9, involved a situation where the lithium layer was not entirely filled with lithium. As such, interactions between the lithium and TM layers had to be taken into account. Again, only the six NN out-of-plane interactions were considered (three from the plane above, three from the plane below). The out-of-plane nearest neighbor lies $c/6$ away in the out-of-plane direction and $a/\sqrt{3}$ away in the in-plane direction such that the distance between out-of-plane nearest-neighbors is roughly 1.004 times larger than for in-plane neighbors based on a = 2.90 Å and c = 14.30 Å as obtained in Chap. 9. Thus, the in-plane and out-of-plane nearest neighbors are nearly equidistant and this correction was included in the calculations.

Chapter 3
Optimization of the Synthesis of Combinatorial Samples

3.1 Experimental Design

Though combinatorial samples were synthesized along each of the single-phase regions I, II, III and IV shown in Fig. 3.1, only the $Li_xNi_{2-x}O_2$ samples (line IV) are reported here in order to describe the mechanisms for lithium loss in the small samples. The methods used to make the samples were closely based on the synthesis done by Carey [10], but several variables were controlled in order to find the conditions that minimize the lithium loss. These included the choice of substrate, the chemical used to cause precipitation (called "precipitator" here), the heating temperature and the atmosphere during heating.

Three substrates were tested: Al_2O_3, MgO and Al_2O_3 treated with LiOH. The main concern with respect to alumina is that it reacts with lithium carbonate to form $LiAlO_2$, a process which has been observed to go to completion at 700 °C [67]. Magnesia shows no such reaction, but it is hygroscopic and porous, thereby requiring large quantities of stearic acid to prevent the water in the solutions from entering the substrate before the co-precipitation reaction takes place. The LiOH treatments were done by spraying 3 M LiOH onto the surface, drying at 55 °C and then heating to 900 °C for an hour in a box furnace. This was repeated three times. Figure 3.2 (d) shows that after the treatments, the substrate had a layer of $LiAlO_2$ (JCPDS #73-1338) on the surface. A fourth treatment typically resulted in cracking and flaking of this layer.

The primary objective here was to synthesize $LiNiO_2$ at 800 °C, the temperature required to make the spinel samples in the Li–Mn–Ni–O system. Since stearic acid has been found to react with lithium to form lithium stearate [68, 69], this chapter also deals with identifying the role of stearic acid in the samples, especially on magnesia where more was required. The precipitators tested were ammonium bicarbonate and ammonium hydroxide, these being the two most commonly used to synthesize metal carbonates or hydroxides from solution. The atmospheres used were either air in a box furnace or oxygen flowing in a tube furnace. The heating temperature was varied

Data in this chapter are reprinted from Ref. [66] with permission from Elsevier.

E. McCalla, *Consequences of Combinatorial Studies of Positive Electrodes for Li-ion Batteries,* Springer Theses, DOI 10.1007/978-3-319-05849-8_3,

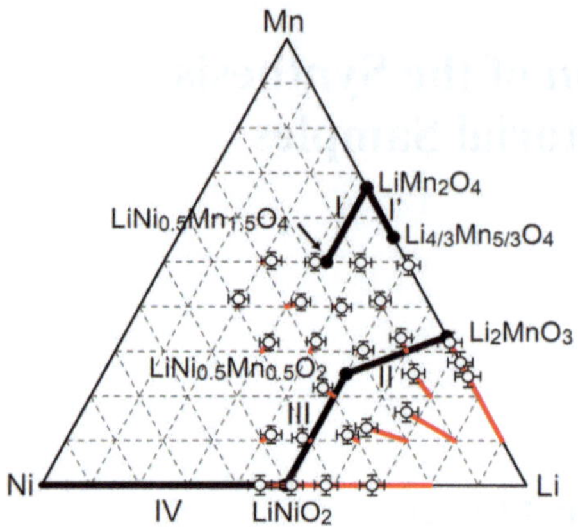

Fig. 3.1 The Li–Mn–Ni oxide pseudo-ternary system where the *corners* refer to the metals used during sample preparation and oxygen content varies throughout the *triangle*. The *bold lines* represent lithium containing single-phase regions known prior to the current project. The *open data points* represent combinatorial samples prepared at 800 °C in oxygen. The compositions were determined by atomic absorption, and the lines leading to each point begin at the as-dispensed compositions; if no line is present the point lies directly above its dispensed composition

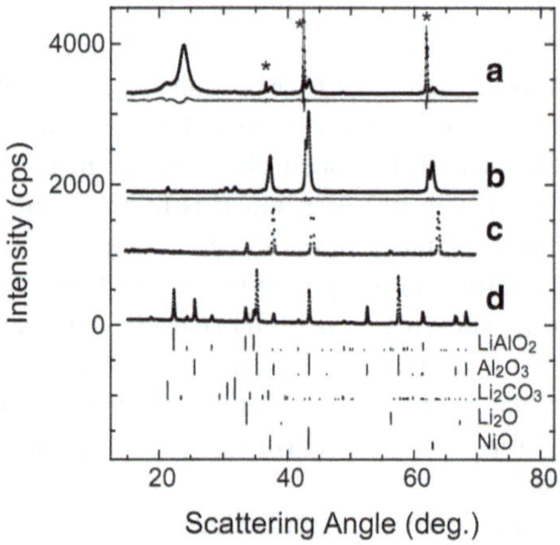

Fig. 3.2 **a** X-Ray diffraction (XRD) scan of a $Li_xNi_{2-x}O_2$ sample with x = 1 as-dispensed with ammonium bicarbonate precipitator, after heating to 400 °C in oxygen. The data are shown with the fit and the difference plot immediately below. The position of the sharp MgO peaks are indicated by *. **b** Sample with x = 1 as dispensed with hydroxide precipitator after heating to 400 °C in air; with fit and difference plot. **c** The bulk sample (x = 0.965) after heating in the thermo-gravimetric analyzer (TGA) in a flow of argon. **d** Al_2O_3 substrate after three treatments with LiOH. *Vertical lines* indicate peaks from the JCPDS database

over the range 200–900 °C to track changes during synthesis. The amount of lithium dispensed was also varied to test the extent to which excess lithium could be used to compensate for the lithium loss.

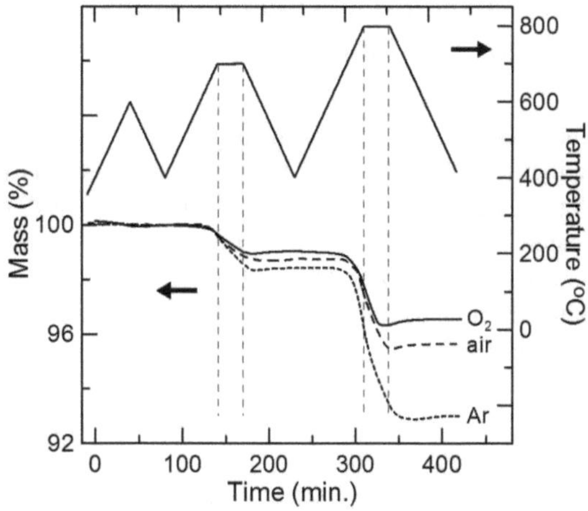

Fig. 3.3 Thermo-Gravimetric analyzer (TGA) data for samples of $Li_{0.965}Ni_{1.035}O_2$ heated in oxygen (*solid line*), air (*dashed line*) and argon (*short dashed line*). A gas flow of 50 mL/min was used in each case. The *vertical dashed lines* mark the start and end of the temperature holds

For the thermo-gravimetric measurements, a bulk $LiNiO_2$ sample of approximately 3.5 g was made by the solid-state reaction of lithium hydroxide (5 % excess) with nickel oxide as described in Sect. 2.1.2. The mixture was placed in an alumina boat and heated to 750 °C for 10 h in air. For each TGA run, 11–13 mg were heated in a TA Instrument, SDT-Q600 thermo-gravimetric analyzer (TGA) under a gas flow of 50 mL/min. The experiment was repeated in argon, air and oxygen. Figure 3.3 shows the temperature profile used. Samples were also heated at a rate of 20 °C/min to either 600, 700 or 800 °C and held there for several hours in either a flow of oxygen or air. This was done in order to estimate the expected lithium loss during the 3 h of heating used to synthesize the combinatorial samples.

Combinatorial samples with a mass of less than 1 mg were made by dispensing a total of 10 μL of 1.78 M solutions with a Cartesian Pixsys solution-processing robot (described in Sect. 2.1.1). To see the effect of sample size, a few were made by dispensing a total of 20 μL. The amounts of lithium and nickel nitrate dispensed were varied in order to make $Li_xNi_{2-x}O_2$ with x = 0, 0.8, 0.9, 1, 1.1 and 1.2 if there was no lithium loss (this is referred to as "x as-dispensed"). Samples with x = 1.2 ($Li_{1.2}Ni_{0.8}O_2$) therefore, represent a 50 % lithium excess compared to nickel content with respect to x = 1.

It was also important to determine whether or not the method developed here results in small lithium loss throughout the entire Gibbs triangle. To do this, a number of 20 μL combinatorial samples containing varying amounts of lithium, nickel and manganese were prepared by heating at 800 °C for 3 h in flowing oxygen. The samples were then dissolved in concentrated acid and the resulting compositions were determined with atomic absorption spectroscopy.

3.2 Thermo-Gravimetric Analyzer (TGA) Results for Lithium Loss During Synthesis

The bulk sample was characterized using a JD-2000 diffractometer and Rietveld refinement to determine that it was single-phase $Li_xNi_{2-x}O_2$ with x = 0.965. Figure 3.2 (c) shows the X-ray diffraction (XRD) scan of this sample after being heated to 900 °C in argon in the TGA. The scan contains two phases: $Li_xNi_{2-x}O_2$ with x < 0.62, and Li_2O (JCPDS #77-2144). Lithium oxide is therefore one product of the thermal decomposition of lithium nickel oxide and balancing the equation with oxygen yields:

$$Li_yNi_{2-y}O_2 \rightarrow aLi_xNi_{2-x}O_2 + b[Li_2O + \frac{1}{2}O_2] \tag{3.1}$$

where $a = (2-y)/(2-x)$, $b = (y-x)/(2-x)$ and $y > x$. This reaction was first proposed by Antolini while studying the thermal decomposition of lithium nickel oxide, $Li_yNi_{2-y}O_2$, with an initial lithium content of y < 0.6 using thermo-gravimetric analysis at temperatures of 900 °C and higher [70].

In order to confirm Eq. 3.1 and to verify that both lithium oxide and oxygen are lost, the X-ray scattering from certain samples was measured after the TGA runs used to produce Fig. 3.3. The sample that was heated in air initially had a mass of 11.66 mg and 0.70 mg was lost during the TGA measurements. Equation 3.1 and the assumption that both O_2 and Li_2O are lost were used to determine that the final state corresponded to x = 0.828. XRD with the JD-2000 diffractometer analyzed with Rietveld refinement yielded x = 0.839. Similarly, the sample heated in oxygen had a lithium content of x = 0.887 according to XRD in good agreement with 0.870 obtained from Eq. 3.1.

Furthermore, Sata found that lithium oxide then reacts with oxygen according to: $2Li_2O_{(s)} + O_{2(g)} \rightarrow 2Li_2O_{2(g)}$ [31]. Thus, the means for lithium loss during synthesis of the combinatorial samples may be due to the production of lithium oxide which is then converted to lithium peroxide vapor. The fact that solid Li_2O was only seen in the XRD scans after heating in argon is consistent with the loss of lithium arising from the reaction of lithium oxide with oxygen identified by Sata [31]. Since lithium oxide does not decompose or evaporate at these temperatures and the TGA data are consistent with losing both oxygen and lithium oxide, the decomposition of lithium nickel oxide followed by the formation of lithium peroxide vapor was the likeliest source of lithium loss in the bulk samples.

Figure 3.3 shows the mass loss of samples heated in argon, air and oxygen. Clearly, the reaction for lithium loss was slowed by oxygen. Furthermore, the sample heated in argon showed continued mass loss during cooling after the hold at 800 °C while the samples heated in oxygen and air showed no such loss and perhaps even a small mass increase during the initial stages of cooling. This effect suggests that oxygen re-entered the sample during cooling in order to counter a deficiency which emerged during heating and only samples in argon continued to lose mass.

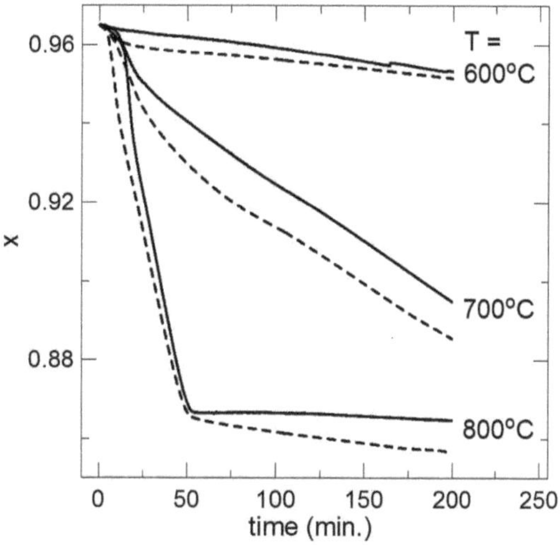

Fig. 3.4 TGA results for holds at 600, 700 and 800 °C in either oxygen (*solid lines*) or air (*dashed lines*). The samples were heated at 20 °C/min and t = 0 min represents the moment when the temperature first reached its intended value. The lithium content was calculated from the mass loss using Eq. 3.1 assuming that both lithium oxide and oxygen are lost

Figure 3.4 shows the lithium content in the samples held at a constant temperature in either oxygen or air. The lithium content is calculated using Eq. 3.1 with $y = 0.965$ and assuming that both lithium oxide and oxygen are lost. Since this does not take into account the oxygen deficiency suggested by Fig. 3.3, the calculated x-value represents a lower limit on the actual lithium content. The samples heated at 800 °C clearly showed two regimes as described by Antolini [70]. His data showed similar results and he attributed the rapid initial lithium loss to the reaction being limited by diffusion along grain boundaries. Next, the rate diminished and this was consistent with the lithium loss being limited by diffusion within grains. The first 50 min seen here at 800 °C in Fig. 3.3 were consistent with the more rapid loss attributed to the reaction being limited by diffusion along grain boundaries, while after this point, the grain surfaces were depleted sufficiently so that the limiting step was diffusion within grains. During the first step, the difference between the behaviors in the two gases was slight. In contrast, at 700 °C, a single regime was seen. At this temperature, the change of rate occurred after approximately 4.5 h, and so the lithium loss was roughly proportional to time at this temperature during the 3 h heating of the combinatorial samples. It is also significant that the lithium loss over 3 h in oxygen was only about 15 % lower than in air. Similarly, only one regime was seen at 600 °C and the mass loss was quite small, particularly in oxygen.

Figure 3.4 suggests that a sample with an initial value of $y = 1$ should end up with a lithium content of approximately $x = 0.94$ after heating for 3 h at 700 °C in air. Similarly, a sample with $y = 1$, heated at 800 °C in oxygen should have a final x value of about 0.90. Combinatorial samples were typically less than 1 mg in mass such that a larger lithium loss was expected for these samples, but the value of 0.90 can be considered as an ideal objective for the combinatorial samples. These data therefore clearly show that excess lithium is required to synthesize combinatorial samples of $LiNiO_2$ at 800 °C.

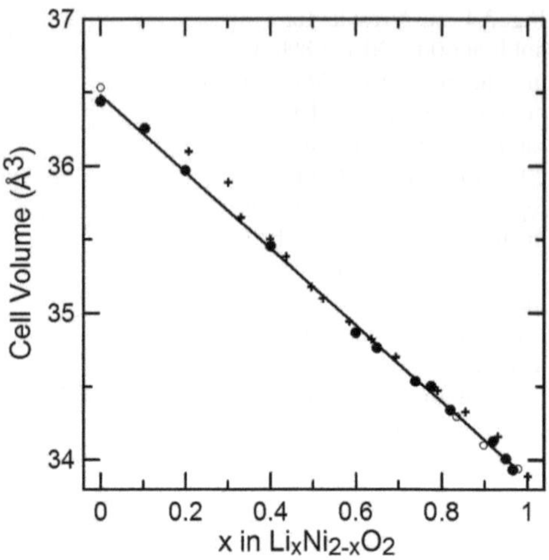

Fig. 3.5 Unit cell volume versus x in $Li_xNi_{2-x}O_2$ obtained by Goodenough et al. [71] (*closed circles*), Li et al. [36] (*crosses*), and from the present work scanned on the JD-2000 and analyzed with Rietveld refinement (*open circles*). The unit cell volume is per $Li_xNi_{2-x}O_2$ formula unit

3.3 X-Ray Diffraction (XRD) Results of Lithium Loss

3.3.1 *Combinatorial Samples*

Accurate values for the lithium content of the samples after heating were obtained from the lattice parameters. The $Li_xNi_{2-x}O_2$ structures are cubic up to $x = 0.62$, but it has been shown that the entire range of samples from $x = 0$ to 1 can be described with hexagonal structure using peak indexing according to $LiNiO_2$, JCPDS #89-3601 [32, 36, 71]. Figure 3.5 shows the cell volume as a function of x as measured by Li [36] and Goodenough [71]. Samples measured with the JD-2000 were also included in the graph since the lattice parameters and lithium content were independent fitting parameters in the Rietveld refinement. The data are linear over the entire compositional range and well defined by $V = 36.479 - 2.6048x$ which is a fit to the Goodenough data only.

Table 3.1 shows the values for x, size and strain obtained for a wide variety of combinatorial samples prepared as described in Sect. 2.1.1. The samples measured on both the JD-2000 and the Bruker show that using the unit cell volume to calculate x gives the same value as obtained by Rietveld refinement within 0.02. Furthermore, the four combinatorial samples that were prepared in oxygen and fit with Rietveld refinement showed that the maximum amount of nickel on the lithium layers was 4.4 $\pm$ 2.1 % more than that expected for a perfectly ordered structure at 800 °C and 3.5 $\pm$ 1.6 % at 700 °C. Although these values are slightly higher than the value of 2.1 % obtained by Li et al. [36], it can be concluded that the amount of nickel disorder on the lithium layers in the combinatorial samples was not greatly different from that seen in bulk samples. Also, the Bragg R-factors obtained for the combinatorial samples ranged over 2.2–3.5 %, compared to 2.3–3.4 % for Li et al. showing that high quality fits were generated by the refinement. When heated in air, the lithium loss is

Table 3.1 $Li_xNi_{2-x}O_2$ properties obtained by fitting the X-ray diffraction (XRD) patterns of the combinatorial samples. All samples were prepared by heating for 3 h in a flow of approximately 30 mL/min of oxygen gas or in air in a box furnace. The uncertainty values shown are the maximum for the entries in each column. Size is shown for the samples made at 700 °C only

Substrate	Precipitator	Gas flow	x initial	x calculated ($\pm$0.036)[c]		size (nm) ($\pm$20 %)	strain (x10^{-3} $\pm$0.004x10^{-3})	
				700 °C	800 °C		700 °C	800 °C
Al_2O_3	CO_3^{2-}	Air	1.0	0.527	0.529[a]	37	0.078	0.019
$LiAlO_2$	CO_3^{2-}	Air	1.0		0.687[a]	71[a]		0.287
MgO	CO_3^{2-}	Air	1.0	0.681		55	0.278	
			1.0	0.878		54	0.388	
				0.898[c]	0.834[c]			
MgO	CO_3^{2-}	O_2	1.09	0.967		89	0.319	
				0.977[c]	0.914[c]			
			1.2	0.994		73	0.176	
			0.9	0.796		61	0.384	
MgO	OH^-	O_2	1.0	0.886		66	0.379	
			1.2	0.987		117	0.174	
			0.8	0.704	0.650	40	0.207	0.293
Al_2O_3	OH^-	O_2	1.0	0.906	0.910	57	0.655	0.118
			1.2	1.01	0.952	133	0.104	0.111
			0.0	0.065	0.052	134	0.077	0.055
$LiAlO_2$	OH^-	O_2	1.0	0.923	0.828	—[b]	0.196	0.032
			1.2	0.938	0.829	111	0.149	0.011

[a] Data collected at 750 °C by Graham Carey
[b] Size values greater than 150 nm are omitted as the size broadening is indistinguishable from the XRD machine peak broadening
[c] Scanned on the JD-2000 diffractometer; fit with Rietveld refinement using the Rietica software.
[d] The uncertainty on x was reduced to 0.006 for x > 0.7

extreme: 47 % on alumina at 700 °C. This is reduced to approximately 32 % by using either magnesia or alumina treated with lithium hydroxide. The poor performance of the alumina substrate implies that it reacted with the lithium in the samples to form $LiAlO_2$. However, even the samples with 50 % excess lithium on magnesia had a low lithium content (x = 0.769) at 700 °C showing that excess lithium is not sufficient to synthesize $LiNiO_2$ in air and that the lithium loss is far greater than the 6 % expected from the TGA data.

Oxygen played a dramatic role in reducing the lithium loss. The best result at 700 °C without using excess lithium was x = 0.923 (< 8 % Li loss) obtained on alumina treated with LiOH, though this is quite close to the value of x = 0.906 obtained on alumina. The reduction of lithium loss in oxygen as compared to air was far greater than the 15 % difference seen in the TGA results. These observations suggest that the thermal decomposition of lithium nickel oxide is not the only mechanism for lithium loss in these samples. Additionally, the substrate had little effect on the lithium content when heated in oxygen, showing that the reaction between alumina and lithium was suppressed.

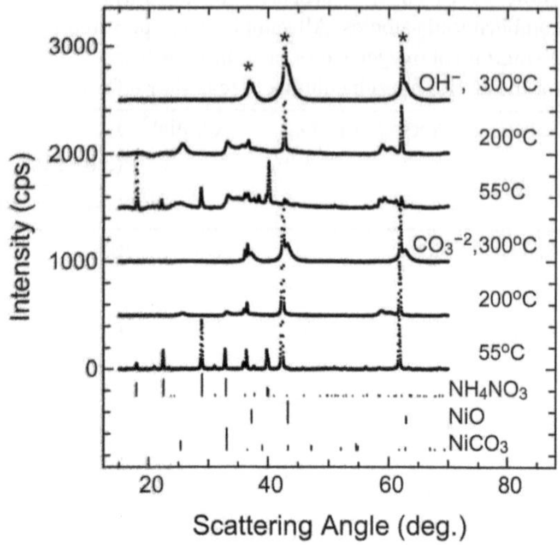

Fig. 3.6 X-Ray diffraction (XRD) scans of nickel nitrate samples obtained after co-precipitation, drying and heating to various temperatures. The MgO peaks were truncated at 500 cps and * indicates their positions. For ease of viewing, a polynomial background was fit and subtracted from each scan. The *vertical lines* indicate peak positions and relative intensities from the JCPDS database

The results were somewhat erratic on $LiAlO_2$, which may be due to inconsistencies in the thickness of the LiOH layers. The sample with x = 0 as-dispensed was used to see if the $LiAlO_2$ layer gave some lithium to the samples. The results of x = 0.06 and 0.05 show that a small amount of lithium was transferred to the nickel oxide but this is not considered to be a significant factor.

The result of x = 0.910 obtained on alumina at 800 °C in oxygen without excess lithium shows that the lithium loss under these conditions was comparable to the ideal situation found with the TGA where a value of x = 0.90 was predicted. Also, the results of x = 0.952 obtained under the same conditions with 50 % excess lithium shows that samples of $Li_xNi_{2-x}O_2$ can be synthesized over nearly the whole x-range under the same conditions as required to prepare the spinel samples. Though a sample with x = 1 was not made, the objective of finding the conditions that minimize the lithium loss was reached and x = 0.95 is comparable to that obtained for bulk samples heated in air. Figure 3.1 shows that samples synthesized throughout the Gibbs triangle under these conditions result in little lithium loss. The only exceptions are samples near the lithium corner where large losses occurred because Li_2O cannot exist in atmospheres containing oxygen as previously discussed.

3.3.2 Combinatorial Samples During Synthesis

To determine the cause of the extra lithium loss in air that was not accounted for by the thermal decomposition of $LiNiO_2$ and to better understand how the combinatorial samples were synthesized, a number of samples were prepared and measured with the Bruker diffractometer after heating to various temperatures. Combinatorial samples were dispensed on magnesia, dried at 55 °C and the XRD spectra were measured directly on the substrate. Figure 3.6 shows the results for two samples prepared from nickel nitrate only. There were a number of phases present in these scans with the strongest peaks being from ammonium nitrate (JCPDS #83-0520), which arose as a product of the reaction between the metal nitrates and the precipitator. The samples were then heated for 3 h at 200 °C and scanned again. As expected, the ammonium nitrate decomposed at this temperature [72], and the peaks due to this phase were not seen beyond this point. The heating step was then repeated at 300, 400 . . . 900 °C and the scattering from each of the samples was measured after each heating step. Figure 3.6 shows that the $x = 0$ samples contained NiO only (JCPDS #47-1049) after being heated at 300 °C such that all contaminants are eliminated at this temperature.

Magnesia was selected for this part of the study because it only has three X-ray peaks in the range over which scans were taken. These peaks were fit to and subtracted from the patterns. In practice, the MgO peaks (JCPDS #45-0946) were shifted from their expected positions because the X-ray machine was aligned to the top of the samples such that the substrate had a vertical misalignment of up to 1 mm. This misalignment varied from sample to sample such that an independent zero in scattering angle was needed for these peaks.

Some samples studied in this way were prepared with the bicarbonate precipitator heated in oxygen, while others were prepared from ammonium hydroxide heated in air. In order to help distinguish the effect of the precipitator from the atmosphere, samples prepared with ammonium hydroxide and heated in oxygen were measured after heating to 600, 700, 800 and 900 °C only. Figure 3.2 (a) and (b) show XRD scans obtained by this approach, along with the fits and difference plots. The lithium hydroxide converted to lithium carbonate, and this occurred readily at 300 °C while the stearic acid decomposed. Even the smaller amounts of stearic acid used on alumina (approximately 1.5 mg/cm^2) were sufficient to produce enough carbon dioxide to react with all the lithium in the combinatorial samples. Thus, the synthesis of lithium nickel oxide in the combinatorial samples involved the reaction of lithium carbonate with nickel oxide regardless of the precipitator used. The fitting of the XRD scans therefore required three phases: lithium carbonate (JCPDS #80-1307), lithium nickel oxide and magnesium oxide.

Figure 3.2 (a) shows that the scattering from samples prepared with ammonium bicarbonate contained a large peak near 23.5°. The position of this peak varied from sample to sample, following the magnesia peaks showing that it can be attributed to the surface of the substrate. Lala et al. [68] found that lithium stearate has a large peak near 24° and a smaller peak above 40°. Since the peak near 24° was only seen in samples with lithium, it was most likely lithium stearate, though there may have been some nickel stearate as well since it also has peaks in this range [73]. The peak was fit using a pseudo-Voigt function. Figure 3.7 shows that the stearate peak grew as the samples were heated, especially during the stearic acid decomposition. The peak disappeared after heating at 600 °C in most samples, consistent with the fact that lithium stearate converts to lithium carbonate at 522 °C [68]. The lithium stearate on the substrate surface results in a lithium deficiency in the sample during heating that could affect the final lithium content of the samples. The stearate peak was never seen when ammonium hydroxide was used as the precipitator. It is also possible that some lithium stearate decomposition also contributed to the lithium loss.

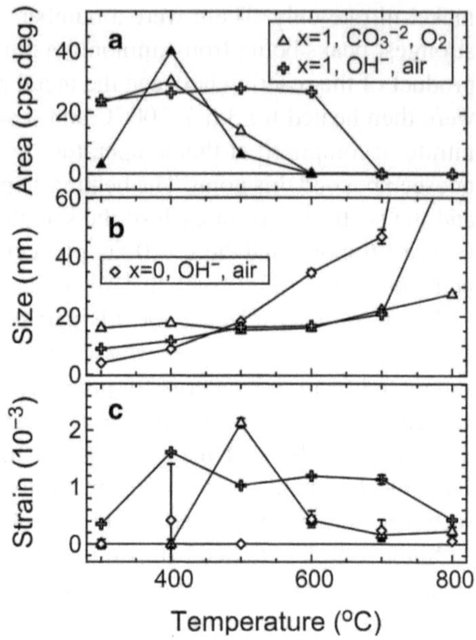

Fig. 3.7 **a** Peak areas as a function of temperature. *Open symbols* are for the Li_2CO_3 (002) peak. The *closed triangles* represent the large stearate peak (divided by 30) obtained with ammonium bicarbonate and heated in air. **b** $Li_xNi_{2-x}O_2$ crystallite size. **c** Strain as a function of temperature. Error bars smaller than the symbols were omitted

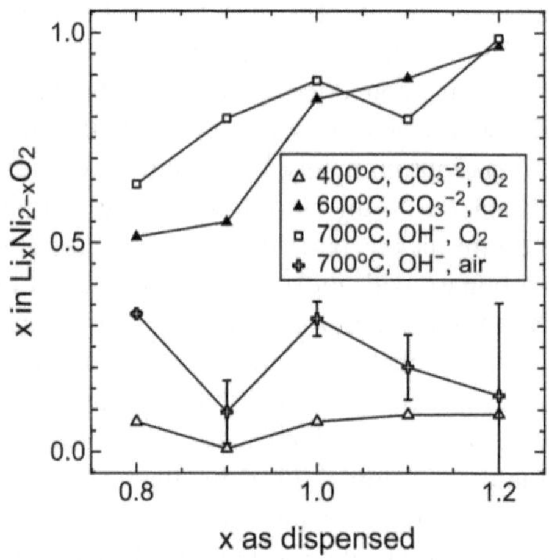

Fig. 3.8 Lithium content measured by X-ray diffraction (XRD) versus the amount dispensed

Figure 3.8 shows that at 400 °C, the amount of lithium in the lithium nickel oxide structure was constant regardless of the amounts dispensed. At this temperature, the amount of lithium carbonate coexisting with the structure was proportional to the amount of lithium dispensed. Since the excess lithium found in the $x = 1.1$ and 1.2 samples did not react into the nickel oxide structure, the lithium nickel oxide synthesis was limited by the slow rate of reaction between lithium carbonate and the rocksalt structure at low temperatures.

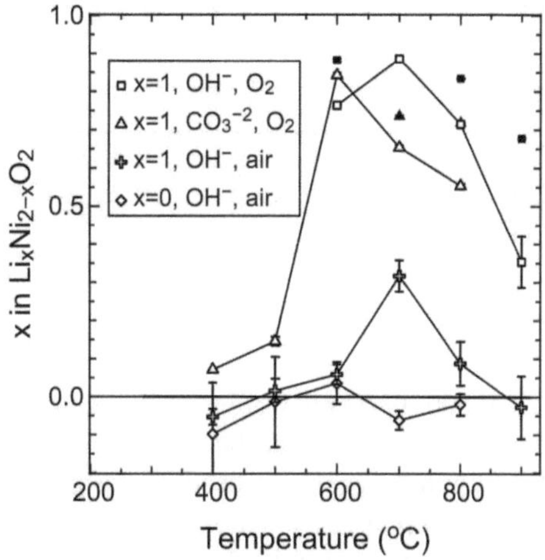

Fig. 3.9 Calculated lithium content as a function of temperature. *Closed symbols* indicate samples made by dispensing 20 μL of solution as compared to 10 μL for the *open symbols*

Figure 3.9 shows the calculated value of x for samples with $x = 0$ and 1 as-dispensed as a function of the heating temperature. Samples with $x = 0.8$, 0.9, 1.1 and 1.2 showed the same trends as the $x = 1$ sample; they were excluded for clarity. The amount of lithium in the rocksalt structure stays low in air with the only significant increase taking place at 700 °C where $x = 0.32$. This behavior is in sharp contrast to that seen in the spinel $LiMn_2O_4$ samples where lithium mixes in readily at 250 °C [69]. Figure 3.7 (a) shows that lithium carbonate coexisted with lithium nickel oxide up to 600 °C in air. Since lithium carbonate decomposes spontaneously at 640 °C [74], there was no temperature in air at which lithium carbonate reacted rapidly with nickel oxide and lithium carbonate did not decompose. Both reactions occurring simultaneously above 640 °C keep the lithium content low throughout the experiment. The lithium loss in these samples is therefore attributed primarily to the slow reaction rate for the formation of $LiNiO_2$ and is not the result of the thermal decomposition of lithium nickel oxide. In a flow of oxygen, lithium carbonate was not present after heating at 600 °C and the lithium content jumped to $x = 0.85$ at this point. This shows that the reaction between lithium carbonate and the NiO structure

$$2Li_2CO_3 + 4NiO + O_2 \rightarrow 4LiNiO_2 + 2CO_2 \tag{3.2}$$

is significantly slower in air and occurs readily at 600 °C in oxygen. This temperature is important because lithium carbonate does not yet decompose. Thus, the second mechanism for lithium loss in the combinatorial samples is the thermal decomposition of lithium carbonate. The fact that the substrate has little effect on the lithium content of the samples in oxygen implies that lithium already in the $Li_xNi_{2-x}O_2$ structure does not react with alumina. It is therefore the accelerated formation of lithium nickel oxide that resulted in the suppression of the formation of $LiAlO_2$ for samples heated on alumina in oxygen.

In order to determine why the formation of lithium nickel oxide is hindered in air, 10 μL combinatorial samples with $x = 1$ as-dispensed were prepared and heated for

Table 3.2 Average Li_2CO_3 crystallite size and microstrain. Values were stable over the heating temperature range in which lithium carbonate is present. The uncertainty is half the range in the values

Precipitator	Atmosphere	Size (nm)	Strain (10^{-3})	Temperature (°C)
OH^-	Air	22 ± 1	0.14 ± 0.04	300, 400, 500, 600
CO_3^{2-}	Air	51 ± 4	0.30 ± 0.01	400 (2 samples)
CO_3^{2-}	O_2	13 ± 4	0.0[a]	300, 400, 500

[a] The fitting software converged to a strain of 0 which indicates that the machine broadening dominated

3 h at 600 °C. Three were heated in a tube furnace in a flow of 5.5 % O_2, 94.5 % N_2, and XRD showed that the lithium content in these samples was $x = 0.91 \pm 0.03$. When pure oxygen was used, the result was $x = 0.95 \pm 0.01$, such that the amount of oxygen played a small role here implying that there must have been constituents other than oxygen and nitrogen that primarily hindered the synthesis of $LiNiO_2$ when heated in air. Other samples were heated in a tube furnace without the caps on the ends of the tube thereby allowing the escape of carbon dioxide produced during the decomposition of the stearic acid. The XRD scans of these samples showed that lithium carbonate was still present and the lithium content in the lithium nickel oxide structures was $x = 0.68 \pm 0.02$. In a closed box furnace, the result was $x = 0.59 \pm 0.02$ and lithium carbonate was also evident in the scans. This shows that the presence of carbon dioxide slowed down the reaction between lithium carbonate and nickel oxide as one would expect since it is a product of the synthesis. However, venting the excess carbon dioxide had a relatively small effect on the lithium content of the samples, suggesting that another gas such as water vapor may also play a role here. Samples made with double the volume of solution (solid symbols in Fig. 3.9) have higher lithium content after heating, with this effect being more pronounced at higher temperatures. These observations reinforce how sensitive these samples were to surface area. Larger dispense volumes do provide other means to reduce the lithium loss, but 20 μL is the practical limit for samples made with the solution-dispensing robot.

Another means of reducing the lithium loss was found by first heating the samples to 600 °C before heating to the desired temperature. To illustrate this, 20 μL samples of the $x = 1.2$ as-dispensed compositions were prepared by heating at 850 °C for 3 h in oxygen. The resulting XRD scan showed $x = 0.83$, while repeating the process with the addition of a 3 h heating step at 600 °C before heating to 850 °C gave a lithium content of 0.88. Though this effect may be diminished at 800 °C, there should still be some benefit to heating at 600 °C in order to ensure that as much lithium as possible enters the nickel oxide structure at a temperature where the thermal decomposition reaction is slow and lithium carbonate does not decompose.

The values for crystallite size and strain of lithium carbonate were found to be very stable over the temperature range in which it was present. Table 3.2 shows the average values obtained in either air or oxygen. The lithium carbonate crystallites were larger and more strained when the sample was prepared in air and this effect was enhanced if the carbonate precipitator was used. Using the carbonate precipitator in air resulted in crystallites that were considerably larger than the rocksalt structure with which it was reacting (Fig. 3.7 (b)). Furthermore, the lithium carbonate crystallites were small with little strain when heated in oxygen.

Figure 3.7 (b) shows that nickel oxide crystallites grew steadily over the entire temperature range and the strain stayed small. By contrast, when lithium was present, crystallites remained small until all the lithium was done reacting. Figure 3.7 (c) shows that in oxygen, the strain increased dramatically while the lithium entered the structure (500–700 °C) indicative of an increase in the number of defects as well as variations in lithium content within each crystallite [62]. In air, the results were similar with the crystallites only growing rapidly once most of the lithium was reacted. The higher strain was also present over a greater temperature range. Figure 3.7 (c) suggests that heating to 800 °C ensures that the crystallites obtained are relatively homogeneous and have few defects.

3.4 Conclusions Regarding Synthesis of Combinatorial Samples

Mechanisms for lithium loss during and after the formation of combinatorial samples of lithium nickel oxide were identified. During synthesis in air, the main source of lithium loss was the decomposition of lithium carbonate that failed to react with the nickel oxide structure. The fact that the formation of $LiNiO_2$ was hindered significantly in air was attributed to the presence of constituents in air other than oxygen and nitrogen, the likeliest candidates being carbon dioxide and water vapor. The second mechanism for lithium loss was the thermal decomposition of $Li_xNi_{2-x}O_2$. TGA was used to confirm that both lithium oxide and oxygen were lost when the samples were heated in either air or oxygen. In both cases, the loss of lithium from the samples was attributed to the conversion of lithium oxide to lithium peroxide vapor. Synthesizing the samples in dry, carbon dioxide free air would therefore result in lithium content very close to that seen in oxygen, the only difference being attributed to a slight increase in the rate of decomposition of lithium nickel oxide as seen in the TGA. Combinatorial samples of $Li_xNi_{2-x}O_2$ cannot be made with $x > 0.77$ at 700 °C in air. In a flow of oxygen, excess lithium was used to react a sufficient amount of lithium into the material to form $Li_{0.95}Ni_{1.05}O_2$ at 800 °C on an alumina substrate. These conditions allow the simultaneous synthesis of the layered and spinel structures in the Li–Mn–Ni–O system.

Alumina, the least desirable substrate in air, performed comparably to the others in oxygen. Magnesia required large amounts of stearic acid in order to bead the solutions and the subsequent decomposition of the stearic acid was found to interfere with the samples. The third substrate tested was alumina treated with lithium hydroxide and it was difficult to treat it uniformly enough to get consistent results. The substrate selected for further combinatorial studies was therefore alumina in a flow of oxygen. Two precipitators were tested: ammonium bicarbonate and ammonium hydroxide. Little difference can be found in the lithium content of the final samples, but only the carbonate precipitator resulted in the formation of lithium stearate on the surface of the substrate.

The map of lithium loss in Fig. 3.1 shows that samples can be prepared throughout the Gibbs triangle with the exception of samples near the Li corner where the formation of lithium peroxide vapor prevents coexistence with lithium oxide. Consequently, the issue of lithium loss in the Li–Mn–Ni oxide pseudo-ternary system is manageable with the methodology developed here and is used throughout the rest of this thesis for combinatorial synthesis. All other Li–Mn–Ni–O samples were therefore heated at 800 °C for 3 h in oxygen. Various cooling rates were used, but the heating rate was always 5 °C/min. The only variation for Li–Co–Mn–O materials was that they were heated in air in a box furnace since it was found that lithium loss was much smaller for these samples as discussed in the next chapter.

Chapter 4
Combinatorial Studies in the Li–Co–Mn–O System

4.1 Experimental Design

Figure 4.1 shows the results of the current project for the Li–Co–Mn oxide pseudo-ternary system obtained in air at 800 °C. The two single-phase regions that lie solely on the Co–Mn line are a tetragonal spinel phase (including $CoMn_2O_4$) and a bixbyite phase (including Mn_2O_3). Once again, there are two solid solution regions of significance to the Li-ion research community: the cubic spinel and layered regions. The coexistence region between them is relatively simple with two-phase coexistence only. The layered single-phase region and the coexistence between the layered structures and Co_3O_4 are the focus of this chapter. The entire spinel region was also determined [76]. Much of the work for the spinel region was done by Colby Brown and evidence for this will be part of his Master's thesis.

Each combinatorial sample, with an approximate mass of 2 mg, was made using the method described in Sect. 2.1.1. Samples were made for 11 compositions, evenly spaced along each of the three lines: $LiCoO_2 - Li_2MnO_3$, $LiCoO_2$ – sample A_4, and Li_2MnO_3– sample A_4 (using the labels in Fig. 4.1). In this system, lithium loss is less severe than in the case of the Li–Mn–Ni–O layered materials such that 3 h heating in air ($P_{O_2} = 0.21$ atm) was possible. In the vast majority of studies found in the literature, the samples were made by cooling from high temperature at roughly 5–10 °C/min, e.g., [18, 23] . However, the cooling rate can have a significant impact on the phases obtained, so one important feature of the current study was to vary the cooling rate to see the effect on the layered region. The objective was to understand how and at which compositions, the layered–layered nano-composites form in order to determine if the phase diagrams produced with combinatorial samples can be used to predict nano-composite formation. Three cooling rates (quenched, regular cooling and slow cooling) were therefore used.

It was found that lithium loss occurred in all samples at 900 °C when nonquenched, as well as in a few samples heated to 800 °C when slow cooled. This loss resulted in samples lying in the spinel–layered coexistence region (Fig. 4.1) such that Co_3O_4

Data in this chapter are reprinted from Ref. [75] with permission from the American Chemical Society.

E. McCalla, *Consequences of Combinatorial Studies of Positive Electrodes for Li-ion Batteries*, Springer Theses, DOI 10.1007/978-3-319-05849-8_4,

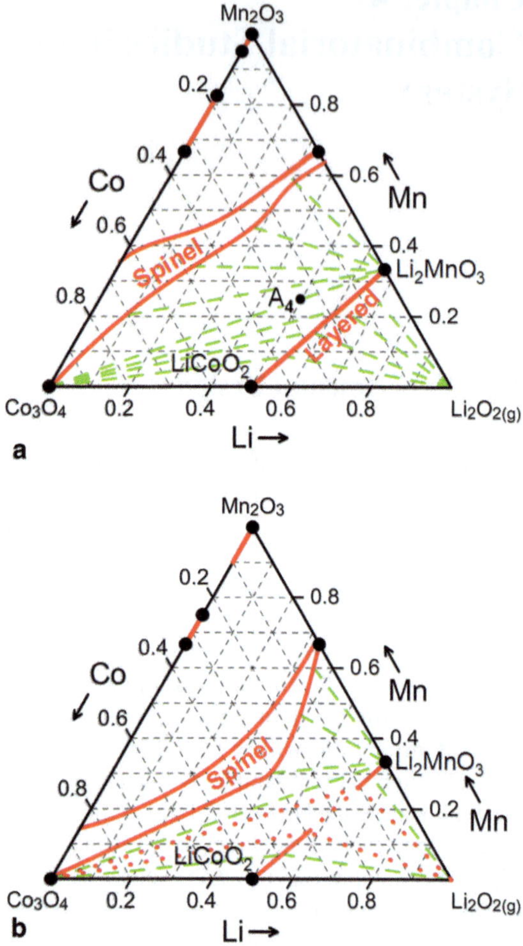

Fig. 4.1 The entire Gibbs triangles for samples heated to 800 °C in air and quenched (**a**), or regular cooled (**b**). Sample A_4 is at the composition $Li_{0.5}Mn_{0.25}Co_{0.25}$ and is referred to throughout the text. *Green dashed lines* are tie-lines, *solid red lines* are boundaries to single-phase regions while *red dotted lines* are tie-lines bounding three-phase regions

peaks could be seen in the X-ray diffraction (XRD) patterns. In order to compensate for this, samples were made along the $LiCoO_2$–Li_2MnO_3 line with 15 % excess lithium. As will be discussed in the results sections below, this eliminated all spinel peaks and and the resulting oxides fell on the desired composition line. There was also no evidence of excess Li_2CO_3 remaining after synthesis, which is another possible product formed when excess lithium does not evaporate as Li_2O_2.

After heating, the samples were characterized using a Bruker D8 Discover X-ray diffractometer. The software written in-house (Sect. 2.3) was used to fit entire X-ray patterns in order to extract lattice parameters and phase compositions. In order to help determine whether the peak broadening seen is consistent with crystallite size effects, the peaks were first corrected for machine broadening and then crystallite size was calculated assuming no microstrain.

To better understand the changes taking place during cooling, the Monte Carlo simulation described in Sect. 2.12 was used with the composition lying midway

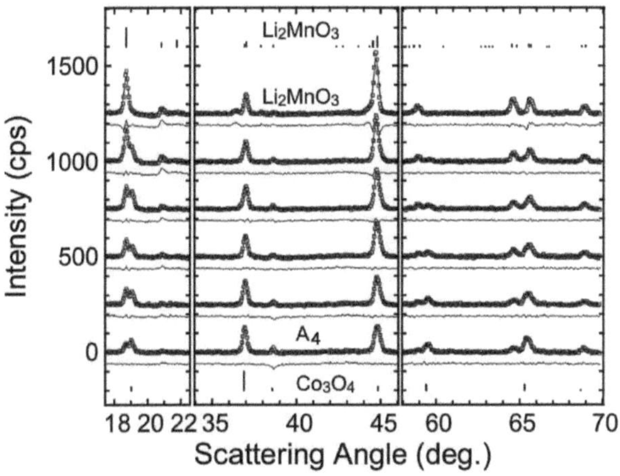

Fig. 4.2 Stack of X-ray diffraction (XRD) patterns with fits and difference plots obtained by heating to 800 °C and regular cooling. The *vertical lines* indicate peaks from the JCPDS database, reference #84-1634 for Li_2MnO_3 and #74-1656 for Co_3O_4. The unlabeled scans between A_4 and Li_2MnO_3 are evenly spaced in composition between the two end members. For clarity, only every third data point is plotted

between $LiCoO_2$ and Li_2MnO_3, i.e., x = 0.5 in $Li[Li_{(1-x)/3}Co_xMn_{(2-2x)/3}]O_2$. The transition metal layer was made up of 50 % cobalt atoms, 33.3 % manganese and 16.7 % lithium, with only lithium on the lithium layers.

4.2 Spinel–Layered Coexistence Region

Figure 4.2 shows a stack of XRD scans taken for 11 samples evenly spaced along the line from Li_2MnO_3 to sample A_4 after heating to 800 °C for 3 h and regular cooling. The scans show two phases: Li_2MnO_3 and Co_3O_4. The peaks at 19.05, 59.5 and 65.2 ° (all indexed to the spinel phase) show most clearly the disappearance of Co_3O_4 for compositions approaching Li_2MnO_3. Figure 4.3 (a) shows a few XRD scans of materials along the line from $LiCoO_2$ to point A_4. Again, coexistence between a layered phase and Co_3O_4 can clearly be seen, though the layered peaks shift to lower angle as the composition moves towards point A_4. This is consistent with the tie-lines shown in Fig. 4.1 (a), which fan out from the cobalt corner such that the lattice parameters of the layered phase increase as the average composition moves from $LiCoO_2$ to A_4 while the spinel lattice parameter remains that of Co_3O_4.

The two-phase scans were fit and this resulted in 20 values of the cubic lattice parameter for the spinel phase for samples along either the $LiCoO_2$–A_4 or A_4–Li_2MnO_3 lines. The average value was a = 8.065 ± 0.002 Å, with a standard deviation of 0.010 Å and is in agreement with the literature value of a = 8.065 Å (JCPDS #74-1656) for Co_3O_4. In contrast, the same 20 compositions, made by quenching from 800 °C, resulted in an average value of a = 8.0761 ± 0.0005 Å with a standard deviation of 0.002 Å. A combinatorial sample made at the cobalt corner by quenching had a lattice parameter of 8.084 ± 0.002 Å, which is in good agreement with the average value obtained in the coexistence region. The larger lattice parameter obtained by quenching is consistent with oxygen vacancies appearing in the spinel structures at high

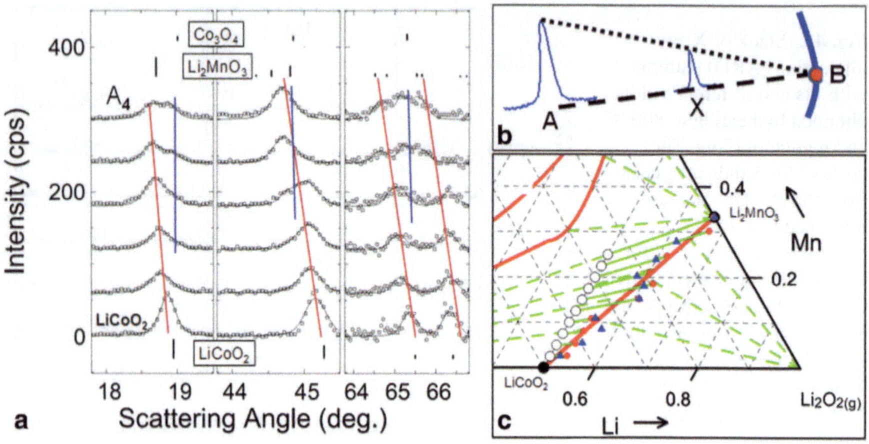

Fig. 4.3 **a** Stack of X-ray diffraction (XRD) patterns for samples lying between $LiCoO_2$ and point A_4 obtained by heating to 800 °C and regular cooling. The *red lines* indicate the layered peaks while the *blue lines* represent the positions of the Co_3O_4 peaks. The *vertical lines* for $LiCoO_2$ are from JCPDS #50-0653. **b** A schematic representation of using the lever rule with a peak of phase A (Co_3O_4 here). The sample *X* lies in the coexistence region such that extrapolating to where the area of the phase *A* peak is zero gives a point *B* on the boundary of the layered region. **c** The results of using the lever rule for samples along the line from A_4 to $LiCoO_2$, represented by the *open circles* (these represent point *X* in **b**). The *red circles* (point *B* in **b**) are the resulting points on the boundary of the layered region for samples made at 800 °C and cooled at the regular rate, while the *blue triangles* were obtained from quenched samples

temperature—a result that was also seen in the spinel structures in the Li–Mn–Ni–O system in Chap. 5. Furthermore, the agreement between the average spinel lattice parameter and that of Co_3O_4 shows that the tie-lines fan-out from the Co corner up to the tie-line connecting Co_3O_4 to Li_2MnO_3 as shown in Fig. 4.1.

In order to determine whether the layered single-phase region is truly a line with no width, the lever rule was used for samples synthesized along the composition line $LiCoO_2$–sample A_4. Figure 4.3 (b) illustrates the use of the lever rule for a sample, X, containing two phases that lie on a tie-line between points A (Co_3O_4 in this case) and B (the layered structure in this example). The fraction of each phase was determined from the XRD fits by dividing the integrated peak area of the largest peak of each phase by the corresponding value for the single-phase sample lying on the same tie-line. The phase fractions were then normalized to add up to unity. The lever rule, described in Sect. 2.4, was used to determine the position of points on the end of the tie-line. Figure 4.3 (c) shows the lever rule results for both quenched and regular cooled samples heated at 800 °C. Within the fluctuations resulting from the experimental methods used, the single-phase layered region is a single line in the Gibbs triangle. This line corresponds to all cobalt atoms being in the 3+ oxidation state while all manganese atoms are in the 4+ state. Generally, Co cannot be oxidized beyond 3+, nor Mn above 4+, under the synthesis conditions used here. The one known exception is that $Li_{1+x}Co_{1-x}O_2$ can be synthesized with x = 0.06 in oxygen [77]. Therefore, there may be a small width to the layered region near $LiCoO_2$, but no evidence for this was seen in the current data.

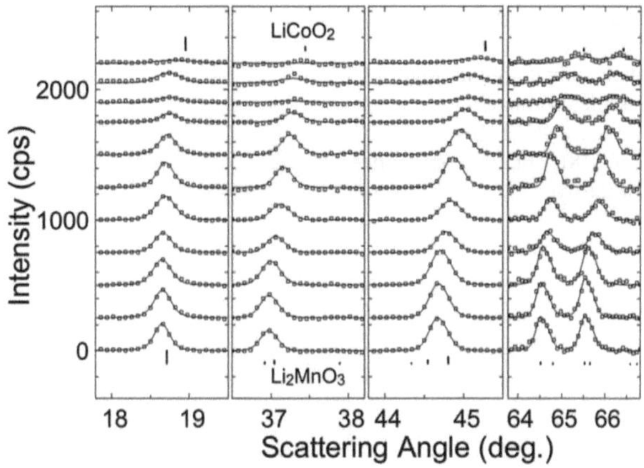

Fig. 4.4 Stack of X-ray diffraction (XRD) patterns with fits for samples made with 15 % excess lithium and heated to 900 °C before quenching to room temperature. For clarity, the second frame (36–38 °) has scattering intensities scaled by a factor of two, while the fourth frame (64–67 °) has a different x-scale and intensities scaled by a factor of five. This same scaling method is used for Figs. 4.7 and 4.9

4.3 $LiCoO_2$–Li_2MnO_3

Figure 4.4 shows expanded regions of the XRD patterns obtained along the $LiCoO_2$–Li_2MnO_3 line by heating to 900 °C and quenching. The scans were consistent with a solid solution over the entire range with no evidence of peak broadening. Similar patterns were obtained by quenching from 800 °C and again samples appeared single phase. Figure 4.5 shows that the resulting fitted lattice parameters are in good agreement with published values from Kim et al. [18] shown in the introduction. In fact, the a lattice parameter follows a straight line over the entire composition range for the samples quenched from 900 °C, while the literature values flatten out for $x > 0.5$. Kim's data obtained by regular cooling from 950 °C more closely resemble the results for quenching from 800 °C. The values for $LiCoO_2$ ($x = 1$) are also very close to the JCPDS values of $a = 2.815$ Å and $c = 14.049$ Å, while the values obtained at the other endpoint, $a = 2.845$ Å and $c = 14.23$ Å, are consistent with Li_2MnO_3 when described with a hexagonal lattice.

Figure 4.6 shows calculated crystallite size assuming no microstrain such that the size parameter was sensitive to any source of peak broadening. All composition lines had duplicates at $x = 0.5$ (if they cannot be distinguished, they are overlapping), such that these reflect the fluctuations in the size values resulting from the small combinatorial samples. The results for quenched samples, both at 800 and 900 °C, show that the size stays roughly linear over the entire composition range except for a sharp increase for Li_2MnO_3 in the data at 800 °C. This implies that at high temperature there is no reduction in crystallite size near the center of the composition line. Ergo, since the high angle peak broadening observed in Kim's data [18] is not seen in any of the quenched samples, must come about during slower cooling.

Figure 4.7 shows the resulting XRD patterns when samples are slow cooled (1 °C/min) from 900 °C. Near each endpoint (i.e., for $x < 0.3$ and $x > 0.7$), the scans appear single phase with the expected shifts in peak position with composition. However, as suggested by the red and blue lines, there is phase separation near the center of the composition line and all peaks are consistent with two layered

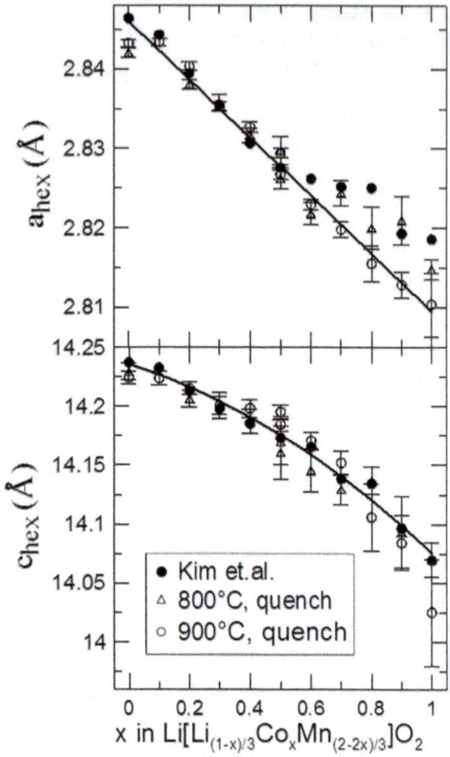

Fig. 4.5 Hexagonal lattice parameters obtained by quenching combinatorial samples compared to literature values from Ref. [18]. The *lines* are guides for the eye

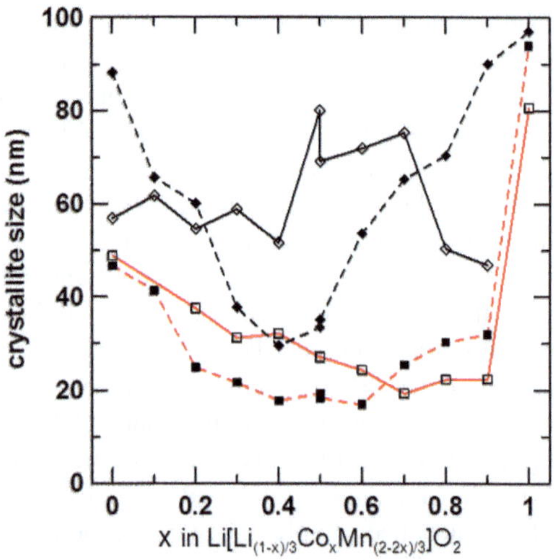

Fig. 4.6 Calculated crystallite size obtained for samples heated at 800 °C (*squares/red lines*), and 900 °C (*diamonds/black lines*). *Open symbols/solid lines* represent samples cooled by quenching, while *closed symbols/dashed lines* are for regular cooling

structures roughly corresponding to the scans at x = 0.2 and 0.7. In order to confirm that lithium loss is not resulting in phase separation here, the region near 19.06 °, where a Co_3O_4 peak (JCPDS #74-1656) would appear, was examined. No Co_3O_4 peaks were present, showing that lithium loss during synthesis was limited to eliminating the excess lithium. Similar peak splitting was observed by Sun et.al. [78] but

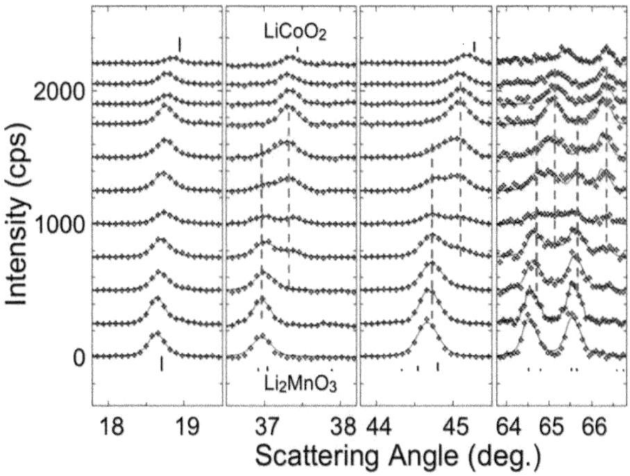

Fig. 4.7 Stack of X-ray diffraction (XRD) patterns, along with two-phase fits, of samples made with 15 % excess lithium and heated to 900 °C before cooling at a rate of 1 °C/min. The *red and blue dashed lines* are guides to the eye in the samples showing two layered structures coexisting

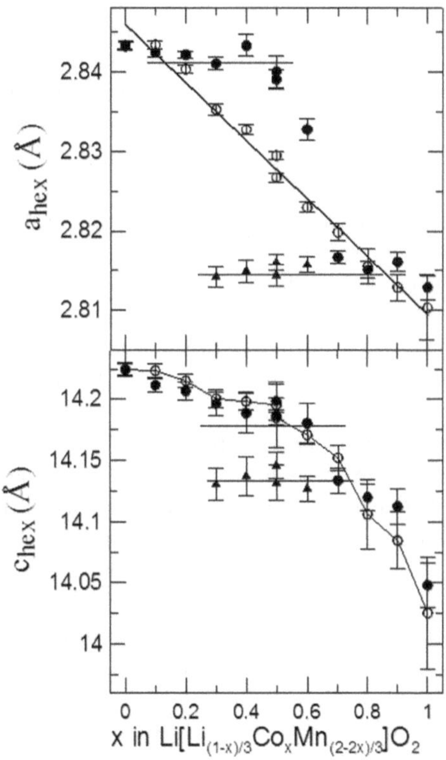

Fig. 4.8 Hexagonal lattice parameters obtained by heating combinatorial samples to 900 °C in air and then either quenching (*open symbols*) or slow cooling (*closed symbols*). All slow cooled samples were fit as two layered structures (near the endpoints, however, one phase always had negligible intensity and so only a single lattice parameter is included for those compositions)

was interpreted as peak splitting due to monoclinic distortions. Tracking these peaks through the entire composition line as shown in Fig. 4.7 clearly shows that the peak splitting is in fact due to the coexistence between two layered phases. Figure 4.8 shows the fitted lattice parameters for samples prepared at 900 °C. Once again, the quenched samples show a single phase only and the lattice parameters progress with

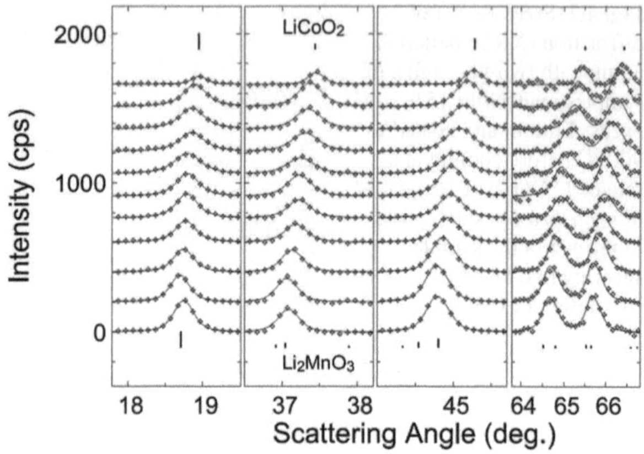

Fig. 4.9 X-ray diffraction (XRD) patterns obtained by regular cooling from 900 °C with 15 % excess lithium. All fits were made assuming a single layered phase

composition as expected from literature. The values obtained by slow cooling show that at both ends, there are solid solution regions, but near the middle, there are two layered phases corresponding roughly to x = 0.2 and 0.8. This confirms that a tie-line exists near the center of the composition line for nonquenched samples. For samples that were slow cooled after heating to 800 °C, the same signs of phase separation were seen: there was extreme peak broadening near the center of the composition line and two samples showed clear peak splitting.

The clear phase separation seen during slow cooling was not seen when regular cooling was used. Figure 4.9 shows the XRD patterns obtained by regular cooling from 900 °C. Excess lithium was used during synthesis because, without excess lithium, two of the twelve samples showed the Co_3O_4 peak at 19.06 °. The scans show that the peaks broaden near the center of the composition range consistent with data from Kim et al. [18]. For example, the peak near 45 ° had a peak width (full width at half maximum) of about 0.26 ° for samples at either endpoint, but that peak broadened to about 0.41 ° at x = 0.4. To quantify this effect and correct for machine broadening, crystallite sizes were once again calculated. Figure 4.6 shows that near each endpoint, crystallite growth continued during regular cooling as one would expect for single-phase materials due to longer time spent at high temperature. In the center of the composition line, the apparent crystallite size diminished. For the samples made at 900 °C this dip corresponds to where phase separation was seen upon slow cooling (1 °C/min) such that the peak broadening seen in the regular cooled samples can be attributed to the onset of phase separation. The same decrease in apparent crystallite size near the center of the line was seen at 800 °C when regular cooled samples were compared to quenched ones, though the change is more subtle. Thus, it can be concluded that regular cooling from 800 °C results in phase separation over the range x = 0.2–0.6, though this may take place over relatively short length scales such that partial XRD peak splitting results in an apparent peak broadening. This phase separation along the layered line gives rise to the three-phase regions shown in Fig. 4.1 (b).

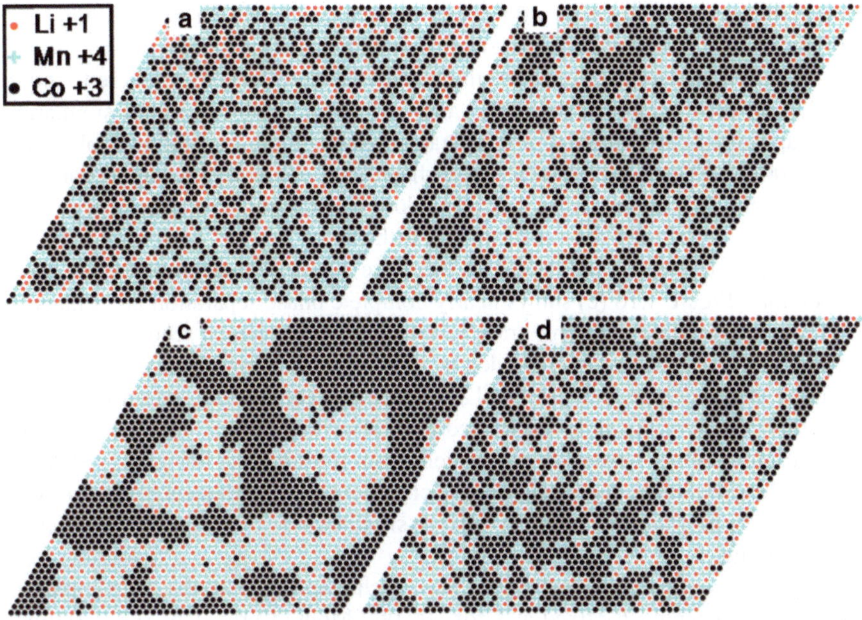

Fig. 4.10 Results of the Monte Carlo simulation for the transition metal layer with x = 0.5 in $Li[Li_{(1-x)/3}Co_xMn_{(2-2x)/3}]O_2$ such that 50 % of the atoms are cobalt, 33.3 % are manganese and 16.7 % are lithium. Results for the simulated slow cool (10000 Monte Carlo steps at each temperature) are: **a** Random occupation of sites (equivalent to $\beta_T = 0$, or infinite temperature). **b** $\beta_T = 1$. **c** $\beta_T = 5$. **d** The result of the simulated anneal at $\beta_T = 1$ (100000 Monte Carlo steps)

As mentioned in the introduction, it is of high interest to have a better understanding of how and at which compositions, layered–layered nano-composites might form in oxide systems. Here, layered–layered phase separation was found to occur on the line between $LiCoO_2$ and Li_2MnO_3, but the composites were not made up of pure $LiCoO_2$ and Li_2MnO_3 as suggested by others [23]. The two phases present contained roughly 20 % disorder wherein one endpoint (x = 0.2) corresponded to 80 % Mn_2Li and 20 % Co on the transition metal (TM) layer, while the other endpoint (x = 0.8) was made up of 80 % Co and 20 % Mn_2Li.

4.4 Monte Carlo Simulation

Figure 4.10 shows results of the Monte Carlo simulation of the material at the center of the layered line. The starting structure shows random occupation of all sites on the hexagonal lattice and therefore represents the product of an instantaneous quench from extremely high temperature. In the simulation, the temperature scale is set by the variable β_T defined in Sect. 2.12. The result for $\beta_T = 1$ corresponds to an intermediate temperature. Figure 4.10 (b) shows signs of phase separation with each domain having some disorder with cobalt present in the Mn_2Li-rich domains and both lithium and manganese in the cobalt-rich domains. This picture illustrates well what was seen experimentally where the layered–layered composites were found to be made up of structures with x = 0.2 and 0.8. As such, $\beta_T = 1$ in the simulation would appear to correspond to a temperature where phase separation took place, below 800 °C. Though not shown, these disordered domains were also found for $\beta_T = 1.5$ and 2.0 demonstrating that these domains minimize the free energy over a range of temperatures during cooling. Furthermore, $\beta_T = 5$ corresponds to 1/5 of the temperature at $\beta_T = 1$, and can therefore be considered to be below room temperature. Figure 4.10 (c) shows nearly complete phase separation takes place at this lower temperature, and only trace amounts of cobalt can be seen in the Mn_2Li domains and these regions are bounded by pure cobalt domains. This picture is very similar to the model made by Bareño [23]. Since this equilibrium state was not seen experimentally here, it can be concluded that slow kinetics must take over at some point during cooling, preventing the system from reaching the low energy state. In terms of the simulation, one would need to consider the activation energy involved in switching two neighboring atoms in order to simulate the "freezing" of the structure during cooling.

In order to confirm that equilibrium was reached in Fig. 4.10 (b), the simulation was repeated, keeping β_T fixed at 1 for 100,000 Monte Carlo steps (i.e., 10 times as many as during the simulated slow cool). Figure 4.10 (d) shows the results of this simulated anneal and confirms that equilibrium was reached in (b) such that the disorder in the two domains cannot be eliminated at this temperature. The domains may also be slightly larger after the anneal: the small cobalt-rich domains in (b) appear to be replaced by a few larger ones in (d). However, the relatively small array makes it unwise to make such generalizations.

The results from the Monte Carlo simulation suggest that $\beta_T = 1$ corresponds roughly to 900 K (i.e., near to but below 800 °C). This implies that the Li–Li nearest neighbor (NN) interaction is $\beta_T k_B T = 78$ meV. Published ab initio calculations of $LiCoO_2$ obtain values of 29 meV as the effective cluster interaction for NN Li–Li clusters, while 7 and 6 meV are obtained for next NN and next next NN, respectively [79]. The interaction energy obtained here is larger; this can be partially attributed to the fact that the only interaction used here is for nearest neighbors. Furthermore, this result implies that a relatively strong interaction was required in order to obtain the phase separation seen during cooling in the current study. One more consequence of this result is that δ, the effective charge fraction of the metal atoms, has a value of 12 % of that expected from the oxidation number. This low value shows that the nearest neighbor interactions are much weaker than that expected from Coulombic interactions assuming all metal–oxygen bonds are purely ionic. It should also be stressed that strain within the lattice was neglected in the simulation (all nearest neighbors were exactly one lattice parameter away from each other). The unit cell volume of $LiCoO_2$ and Li_2MnO_3 differ by 3.4 % while those for x = 0.2 and 0.8 differ by only 1.3 % (using lattice parameters from Ref. [18]). Phase separating over the entire composition line might require fracturing the crystallites, which may be the reason why the endpoints found experimentally were nearer to the center of the line.

4.5 Conclusions Regarding the Formation of Layered–Layered Composites in the Li–Co–Mn–O System

The layered single-phase region in the Li–Co–Mn oxide pseudo-ternary system was explored by a solution-based combinatorial approach. The results showed that the layered region is a single composition line corresponding to cobalt being constrained to the 3+ oxidation state only. This composition line, joining $LiCoO_2$ to Li_2MnO_3, was a solid solution over its entire length when samples were quenched from either 800 or 900 °C. Upon slow cooling, the structures phase separated near the center of the line with the maximum phase separation occurring over the range x = 0.2–0.8 in $Li[Li_{(1-x)/3}Co_xMn_{(2-2x)/3}]O_2$ when samples were cooled from 900 °C at a rate of 1 °C/min. These endpoints correspond to Co and Mn_2Li domains with approximately 20 % disorder on the transition metal layers. Such disorder was also found over a range of temperatures during cooling using a Monte Carlo simulation.

The phase separation upon slow cooling helps explain results from previous studies [18] performed with an intermediate cooling rate where peak broadening seen in the XRD patterns can now be attributed to phase separation on the 2–10 nm length scale with both domains lying on the same lattice as shown in Ref. [22]. The results here show that the nano-scale phase separation occurs when the system has insufficient time to make large scale crystallites of each phase, and the unit cells are close enough in size that the lattice does not fracture upon phase separation. It was also demonstrated that this partial phase separation can be detected by way of careful peak width analysis in the XRD patterns. As such, nano-domain composites can be expected when samples are regular cooled at composition points on the phase diagram that are single-phase when quenched and show layered–layered phase separation when slow cooled. This condition will be used in Chap. 9 to determine the location of nano-composites in the Li–Ni–Mn oxide system where there continues to be considerable debate regarding the structures of the Li-rich layered materials [2, 43–46].

Chapter 5
Combinatorial Studies of the Spinel and Rocksalt Regions in the Li–Mn–Ni–O System

5.1 Experimental Design

In order to map out the entire Li–Mn–Ni–O system, combinatorial samples were made at over 300 different compositions using the method described in Sect. 2.1.1. Figure 5.1 shows these compositions. Due to the large amount of data used in producing the phase diagrams, the supporting evidence is split between two chapters. Figure 5.2 shows the phase diagrams obtained in oxygen. In this chapter, evidence for the spinel and rocksalt phases will be presented. The two-phase regions between Mn_2O_3 and the spinel structures as well as between the spinel and rocksalt materials will also be examined here.

In the lithium loss study described in Chap. 3, two different precipitators were used to cause coprecipitation of the Li, Mn, and Ni mixed nitrate solutions: ammonium bicarbonate and ammonium hydroxide. Although ammonium bicarbonate precipitates all three metals and the hydroxide only precipitates the Mn and Ni atoms, no significant differences were obtained by using either precipitator in the lithium loss study, or in this study. As such, no distinctions will be made in this chapter between the two precipitators, as in all cases, the phases obtained were identical suggesting that ionic transport at high temperature overcomes any lithium inhomogeneity in the starting material.

5.2 Spinel Single-Phase Region

Figure 5.2a shows the phase diagram obtained in this study when samples were heated to 800 °C in an oxygen flow and then quenched to room temperature. Figure 5.2 b shows the phase diagram obtained under the same conditions with the regular cooling rate. Figure 5.3 shows the partial phase diagram obtained by heating to 800 °C in air with regular cooling.

Data in this chapter are reprinted from Ref. [80] with permission from Elsevier.

E. McCalla, *Consequences of Combinatorial Studies of Positive Electrodes for Li-ion Batteries*, Springer Theses, DOI 10.1007/978-3-319-05849-8_5,

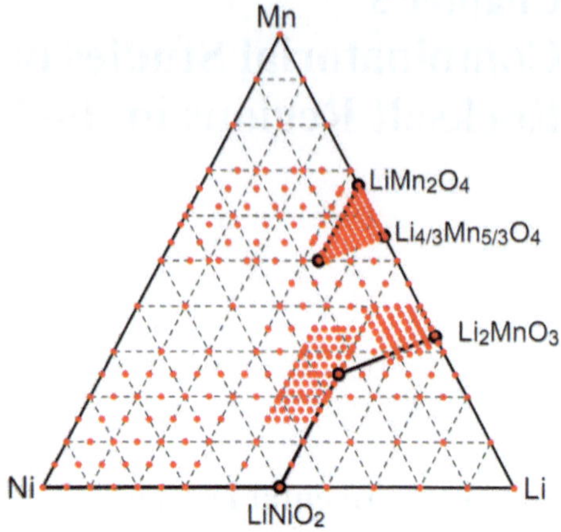

Fig. 5.1 The phase diagram with *red points* indicating all compositions synthesized during the combinatorial studies for regular cooled samples heated in oxygen. Quenched samples were also made at these compositions, with minor variations. The axes are *Li*, *Mn*, and *Ni* metal molar fractions

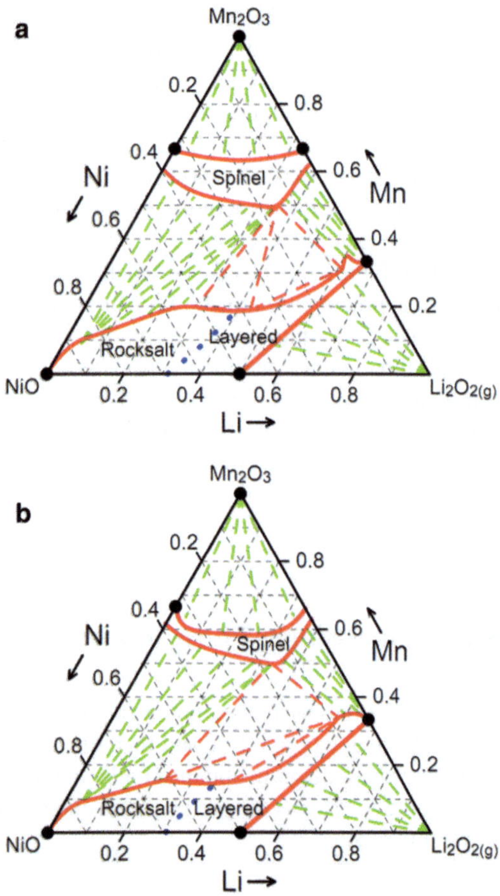

Fig. 5.2 **a** The complete phase diagram obtained by quenching from 800 °C after heating in oxygen for 3 h. The *red lines* are boundaries to the single-phase regions, *green dashed lines* are tie-lines while *red dashed lines* are tie-lines at the outer edges of the three-phase regions. The *blue dotted line* represents a phase transition from the cubic rocksalt to the layered rocksalt structures. **b** The phase diagram for samples heated in flowing oxygen obtained by regular cooling from 800 °C

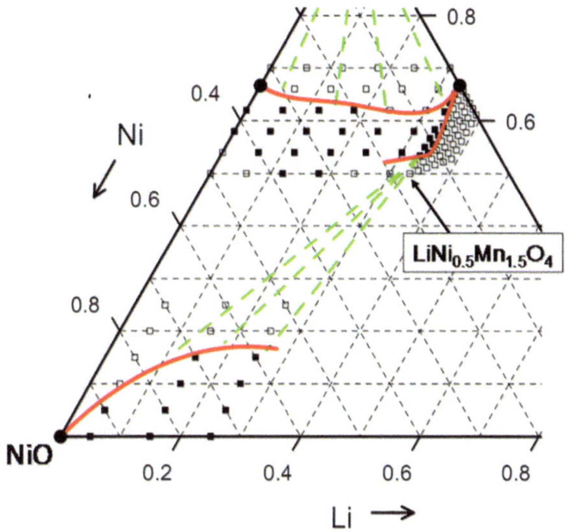

Fig. 5.3 A partial phase diagram obtained in air with regular cooling. Regions showing severe lithium loss were avoided. *Filled points* represent single-phase samples while *empty points* show coexistence of multiple phases

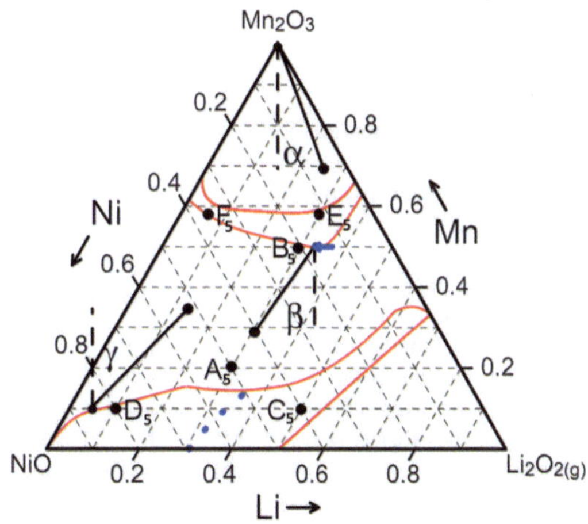

Fig. 5.4 The Gibbs triangle with labels used throughout this chapter

Figure 5.4 shows labels used throughout the results section. A_5–F_5 are specific compositions, while α, β, and γ are angles formed by points in the coexistence regions. Figures 5.5 and 5.6 (left) show X-ray diffraction (XRD) patterns of samples in the spinel region that are evenly spaced between points E_5 and F_5 in Fig. 5.4. The shifting of the (400) spinel Bragg peak near 44° indicates a solid solution. The (220) peak, near 30° in Fig. 5.5, increases in intensity as the composition approaches the Ni–Mn line moving from E_5 to F_5 in Fig. 5.4. Due to the importance of the spinel samples as potential electrode materials, a large number of samples were made in the triangle joining $LiMn_2O_4$, $LiNi_{0.5}Mn_{1.5}O_4$ and $Li_{4/3}Mn_{5/3}O_4$. The samples made in this region of the phase diagram can be seen as a triangle of densely packed points in Fig. 5.1a. Thc XRD scans from these samples made it possible to visually determine

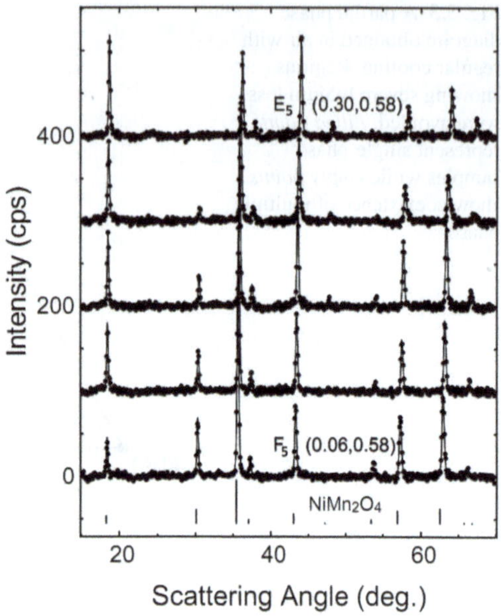

Fig. 5.5 X-ray diffraction (XRD) patterns of a few single-phase spinel structures. Samples E_5 and F_5 are labeled in Fig. 5.4. The database peaks shown are for spinel $NiMn_2O_4$, from JCPDS #84-0542. For clarity, only every third data point is shown

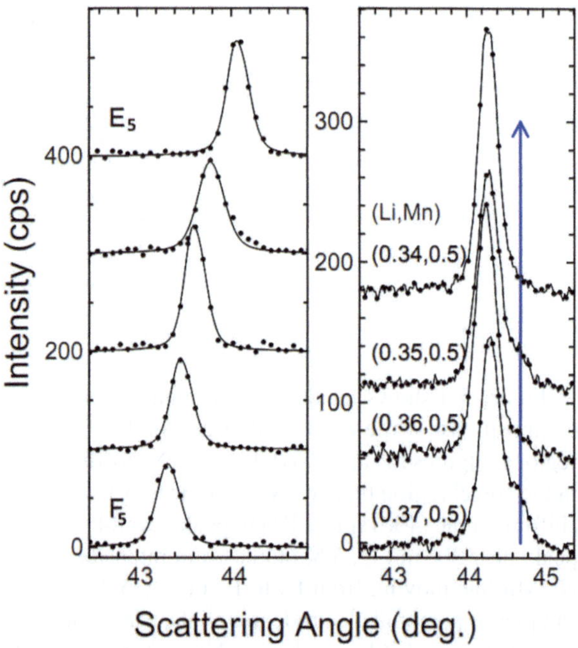

Fig. 5.6 X-ray diffraction (XRD) patterns obtained for single-phase spinel structures. Points E_5 and F_5 are labeled in Fig. 5.4 (*left*). XRD patterns near the boundary of the spinel region; the *blue arrow* matches that shown in Fig. 5.4 and indicates the position of the layered (104) peak here. The other peak is the spinel (400) (*right*)

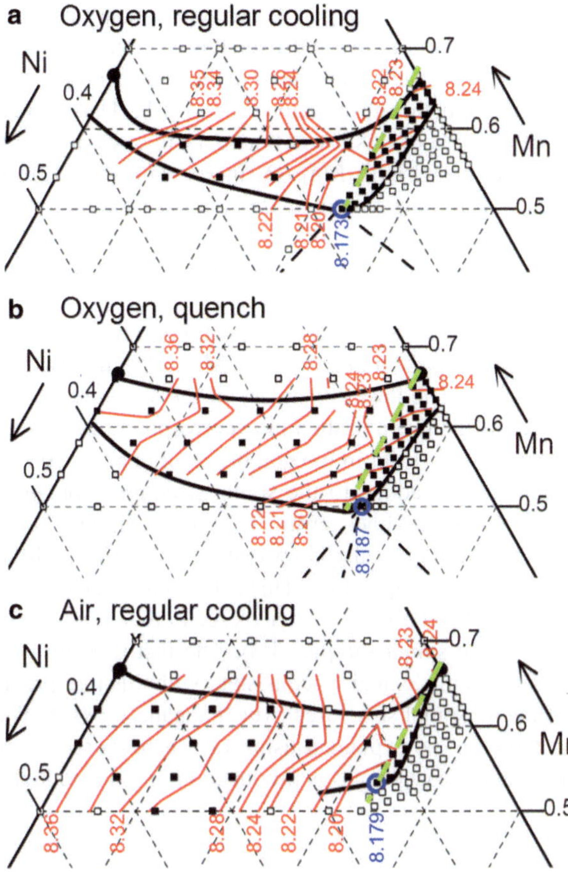

Fig. 5.7 Contour plots of the lattice parameter in the cubic spinel region. *Filled points* indicate single-phase samples while *open symbols* are used to denote coexistence of multiple phases. The *green dashed line* represents the stoichiometric spinel composition line: $LiNi_xMn_{2-x}O_4$; $x \leq 0.5$

where the spinel single phase region ends. Figure 5.6 (right) shows a stack of XRD patterns of samples along the blue arrow in Fig. 5.4. These samples lie on either side of the boundary of the spinel region and the disappearance of the layered (104) peak is used to identify the position of the boundary. Although determining the exact point where the layered peak disappears is difficult, this example illustrates that the error in this method is no more than 0.01 in lithium content (i.e., an error of no more than one scan in the stack).

Figure 5.7 shows contour plots obtained for the spinel lattice parameter throughout the single-phase region. The lattice constants were obtained by whole pattern fitting with the in-house software mentioned in Sect. 2.3. Here, this meant at least eight peaks were used to refine the lattice parameters. The contours extend beyond the single-phase boundary for clarity in labeling, but have no meaning outside the boundaries. Samples synthesized by regular cooling, be it in oxygen or air, show lattice parameters as expected from Zhong et al. for the stoichiometric spinel line $LiNi_xMn_{2-x}O_4$; $x \leq 0.5$ [34]. The consistency between the lattice parameters obtained in air and oxygen when regular cooled extends to the rest of the solid solution region. However, in oxygen, the lower boundary has moved down. This implies that the spinel region is favored over the layered materials as oxygen partial pressure increases. This can be attributed to the higher oxygen content in the spinel structures as discussed in the introduction (Sect. 1.4.1). The cooling rate seems to have no effect on the position of this lower boundary in oxygen even though the minimum lattice parameter near $LiNi_{0.5}Mn_{1.5}O_4$ increases from 8.173 (regular cooling) to 8.187 Å (quenching). Since this increase is not associated with any phase separation, it can be attributed to oxygen deficiencies arising during heating at 800 °C and then being relieved by oxygen returning into the sample during regular cooling, as observed by Ma [37].

Figure 5.7 also shows that excess lithium can be added to the stoichiometric spinel samples. The stoichiometric $LiNi_xMn_{2-x}O_4$ line was included in Fig. 5.7 as a green dashed line. The lower boundary of the single-phase spinel region lies to the right and below the green dashed line, corresponding to an excess of lithium. This effect was more significant in oxygen but even in air at 800 °C, a small amount of excess lithium can be added to the spinel structures containing nickel. Therefore, some spinel–layered tie-lines terminate at points slightly below the $LiNi_xMn_{2-x}O_4$ line.

The contour plots also show that the upper spinel phase boundary moves up significantly at high temperature. This can be interpreted as oxygen loss since coexistence with Mn_2O_3 requires a higher oxygen content per metal atom (3:2) than that of a single-phase spinel sample (4:3). The fact that the upper boundary is also higher for samples made in air is consistent with this interpretation: the lower oxygen content favors spinel structures over manganese oxide. The motion of the boundaries can therefore be explained using Le Chatelier's principle and the fact that at high temperature, oxygen gas is produced giving rise to high entropy.

5.3 Rocksalt Single-Phase Region

Figure 5.8 shows XRD patterns obtained along the C_5–D_5 line in Fig. 5.4. The right panel showing the region near 44° suggests a solid solution. For the samples with 0.4 or 0.5 lithium fraction, all peaks index to the layered structure of $LiNiO_2$ (JCPDS #89-3601). For lower lithium content, the structures become cubic and will be discussed in more detail below.

The location of the phase transition from the cubic to hexagonal layered structures was obtained by fitting all samples as layered and then using the c/a ratio as discussed in Sect. 1.4. Figure 5.9 shows the c/a ratio as a function of lithium content along three lines with various manganese contents. In a cubic structure, $c/a = \sqrt{24}$ such that extrapolating to this value gives the composition where the structures convert from a cubic to a hexagonal structure. Li et al. found that this transition occurs at a lithium content of 0.31 along the Li–Ni line [36]. The blue dotted lines in Figs. 5.2, 5.4 and 5.10 show the position of the phase transition obtained by using the four points generated by this method.

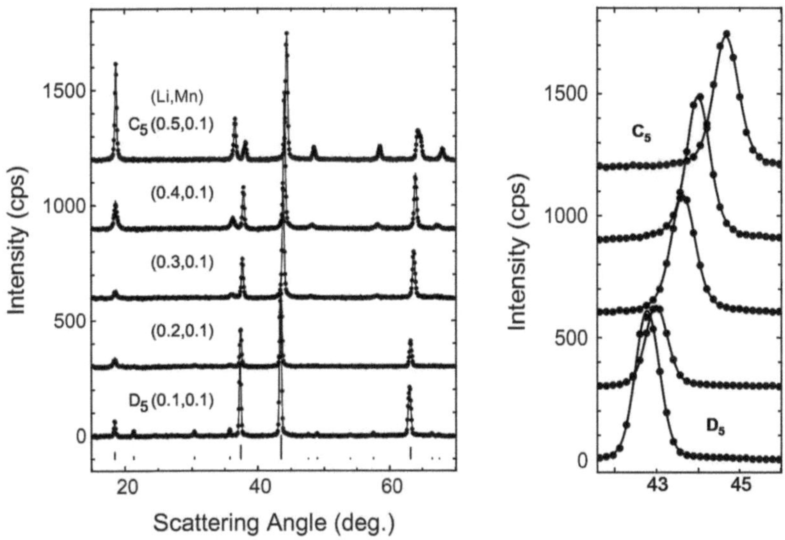

Fig. 5.8 Stack of X-ray diffraction (XRD) patterns for single-phase structures obtained by regular cooling in oxygen (*left panel*). The same scans lines in the *left panel* are for Ni_6MnO_8 from JCPDS reference # 89-4619. C_5 and D_5 refer to compositions defined in Fig. 5.4

The three largest peaks seen in the XRD scan of sample D_5 in Fig. 5.8 can be indexed to $Li_xNi_{1-x}O$; $x < 0.31$, a cubic rocksalt material. However, there are extra peaks present which match up well with the XRD scattering for Ni_6MnO_8. Therefore, sample D_5 shows extra ordering peaks as compared to those expected for rocksalt. Ni_6MnO_8 and Mg_6MnO_8 are isostructural. Kasper and Prener [81] determined that Mg_6MnO_8 takes the Fm3m space group with the following sites:

- 4a sites occupied by Mn at (0, 0, 0)
- 4b sites at (0.5, 0.5, 0.5) are vacant
- 8c oxygen sites at (0.25, 0.25, 0.25)
- 24d sites occupied by Ni at (0, 0.25, 0.25)
- 24e oxygen sites at (x, 0, 0).

To better understand the structure of samples such as D_5 in Fig. 5.8, a regular cooled combinatorial sample synthesized under oxygen at (Li, Mn) = (0.25, 0.15) was scanned using the JD-2000 diffractometer and the pattern was fit using Rietveld refinement. The refinement allowed for manganese and nickel on the 4a sites, all three metals on the 24d sites, and both lithium and nickel on the 4b sites. A key variable tested was the occupation of the 4b sites which are vacant in Ni_6MnO_8. Table 5.1 shows the results for the quality parameters obtained as the 4b occupation changes. Decreasing the occupation below 60 % gave increasingly poorer fits. The best result was obtained with 30 % of the 4b sites vacant. However, it is important to notice that the weak scattering from lithium atoms makes it difficult to distinguish them from

Fig. 5.9 The c/a lattice parameter ratio as a function of Li fraction for three different manganese fractions obtained by regular cooling in oxygen and fitting the X-ray diffraction (XRD) patterns as hexagonal structures

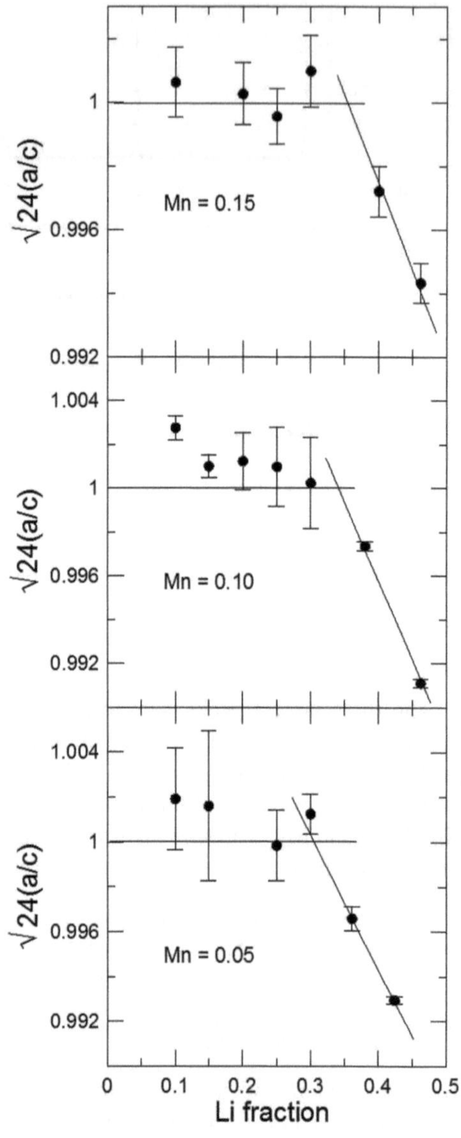

vacancies and so the minimum seen in the quality parameters is shallow. Nonetheless, it is clear that the 4b sites are partially vacant. At this composition, if nickel is in the 2+ state and manganese in the 4+ state, we would expect approximately 19.5 % vacancies on the 4b sites such that the composition would be $Li_{0.244}Mn_{0.146}Ni_{0.585}O$. This is a likely structure, but more study is required to clearly distinguish the lithium atoms from vacancies.

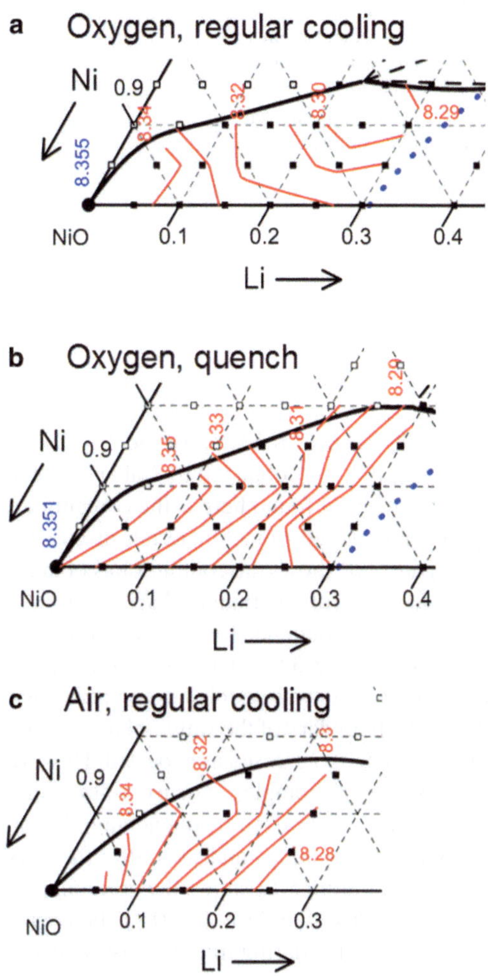

Fig. 5.10 Contour plots of the cubic lattice parameter in the rocksalt region. *Open symbols* show phase separation while the *closed symbols* represent single-phase samples

Table 5.1 Results for the Rietveld refinement of the sample synthesized at the metal composition: $Li_{0.25}Mn_{0.15}Ni_{0.6}$ in oxygen with regular cooling

4b site occupation	R_B (%)	R_P (%)	R_{WP} (%)
100 %	2.76	14.01	19.17
90 %	2.13	13.92	19.09
80 %	1.39	13.84	19.06
70 %	1.18	13.83	19.06
60 %	1.43	13.89	19.07

Table 5.2 shows the refinement results obtained when 30 % of the 4b sites are left vacant. The 4a sites being completely occupied by manganese means that Mn orders on a 2 × 2 × 2 rocksalt cubic lattice making up one-eighth of the metal atoms. The 4b sites form another 2 × 2 × 2 lattice and here lithium orders, with no nickel. This structure will be referred to as ordered rocksalt throughout this thesis. The lattice

Table 5.2 Output from the Rietveld refinement of the sample synthesized at the metal composition: $Li_{0.25}Mn_{0.15}Ni_{0.6}$ in oxygen with regular cooling

Lattice parameter	
$a = 8.304(1)$ Å	
Site occupancies	
4a:	4.0 Mn
4b:	2.8 Li
	1.2 □
8c:	8 O
24d:	4.9 Li
	0.62 Mn
	18.48 Ni
24e:	24 O
	(0.2344,0,0)
$R_B = 1.18\,\%$	

parameter value of 8.304 Å for the 2 × 2 × 2 lattice is in excellent agreement with the value of 4.15 Å obtained by Cabana et al. [38] for the rocksalt lattice confirming that the material found here is the contaminant phase often found when making the spinel and layered–spinel materials discussed in Sect. 1.4.

Figure 5.10 shows the contour plots obtained for the cubic lattice parameter in the rocksalt region of the triangle. As a confirmation that these contours are accurate, the value of 8.33 Å obtained for the contour line through $Li_{0.1}Ni_{0.9}$ in air can be compared to the value of 8.3278 Å obtained for $Li_{0.104}Ni_{0.896}O$ from JCPDS #89-3605 (the lattice parameter was doubled to compare it with the unit cell of the ordered rocksalt). Along the boundary of the single-phase region, the lattice parameter decreased as the lithium content increased as expected. However, the behavior along constant lithium lines was more complex. For example, along the line with 0.2 lithium content in the samples synthesized by quenching, the lattice parameter first increased as more manganese was added and then after the kink near Mn = 0.1, the lattice parameter decreased. Figure 5.11 shows the ordering peaks obtained for samples in this region. In the quenched sample at (0.2, 0.1), there was the first sign of the ordering peaks and so the onset of ordering might be associated with the decrease in lattice parameter. The kink in the contour plot may therefore indicate the location of the order–disorder transition between the rocksalt and ordered rocksalt structures. By contrast, in the regular cooled samples, the ordering peaks were sharp at (0.2, 0.1) and there was even a trace of the peak at 18.5° in the sample at (0.2, 0.05). Here, the contour plot in Fig. 5.10a shows that the lattice parameter decreases steadily as manganese was added, again suggesting a correlation between ordered rocksalt formation and the decrease in lattice parameter. This decrease may therefore be attributed to the ordering of manganese on the cubic lattice resulting in a more efficient packing and thus a decrease in lattice parameter. No transition between the rocksalt and ordered rocksalt was shown on the phase diagrams here, but these preliminary results suggest that there exists an order–disorder transition and that it is sensitive to synthesis conditions. Finding the maximum in lattice parameter as a function of manganese content with a greater number of samples in this region might be one way to determine the position of this phase transition.

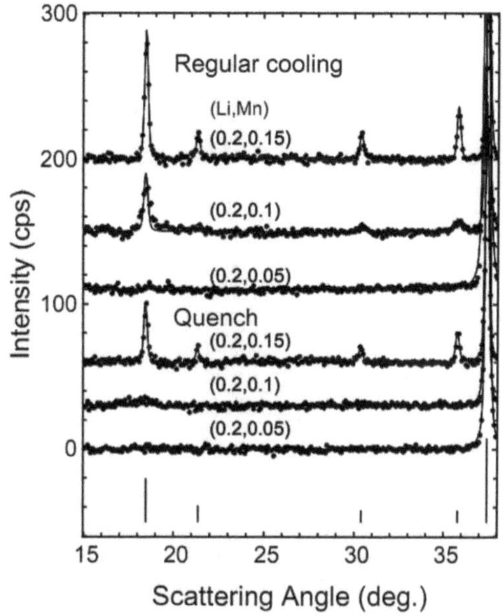

Fig. 5.11 Ordered rocksalt phases obtained in oxygen along the line Li = 0.2. Ordering is more pronounced in the slower cooled samples. The expected peaks for Ni_6MnO_8 (JCPDS #89-4619) are shown as *vertical lines*. The coordinates are (Li, Mn)

Under the conditions used here, Ni_6MnO_8 was present in XRD scans along the Ni–Mn line near the Ni corner showing that it was stable at 800 °C in oxygen. Along the rest of this line, a combination of spinel structures and $NiMnO_3$ was found. However, no lithium containing samples were found to coexist with these structures, so they are of little interest in the context of Li-ion battery research.

5.4 Mn_2O_3–Spinel Coexistence Region

In the manganese oxide–spinel coexistence region at the top of the triangle, XRD scans from 22 samples were fit. The fits were made using the in-house software. The two-phase fitting was performed by keeping the peak width parameters constant at the values obtained for single-phase samples, thereby ensuring that the fitting algorithm did not broaden peaks instead of fitting them with overlapping peaks from two or more different phases. The resulting Mn_2O_3 cubic lattice parameters had an average value of 9.406 ± 0.002 Å and the standard deviation was 0.011 Å. The lattice parameter obtained for the sample synthesized at the Mn corner was 9.409 ± 0.001 Å in excellent agreement with the value of 9.4091 Å given in the JCPDS entry #41-1442 for Mn_2O_3. This result shows that the lattice parameters obtained in the coexistence regions were precise, though the noise on any given measurement was significant and statistics over multiple samples were required. The standard deviation is 0.12 % of the lattice parameter and this value will be used in this chapter and the next as the benchmark to identify a phase from which tie-lines fan out.

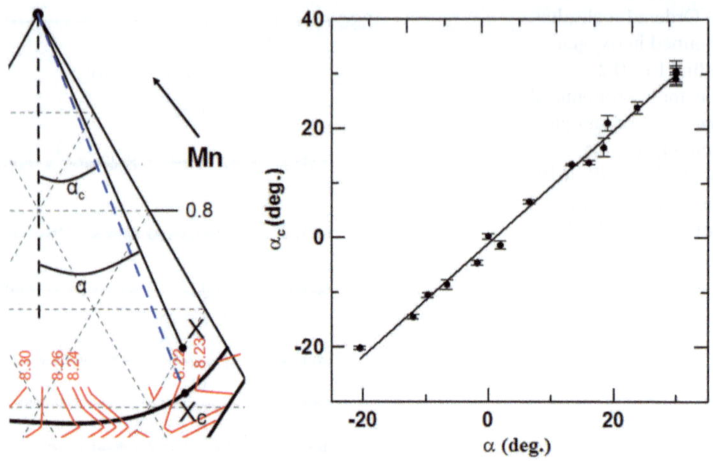

Fig. 5.12 A comparison of the angles α and α_c obtained for a point X, (0.30, 0.66). The spinel lattice parameter at point X was 8.225 ± 0.003 Å such that extrapolation between lattice parameters along the boundary defines point X_c (*left*). α_c as a function of α for samples throughout the coexistence region. The line, $\alpha_c = 1.03\alpha - 1.2$ is a linear fit to the data (*right*)

In this coexistence region, the tie-lines were easy to determine since they must fan out from the manganese oxide corner. However, it was important to establish that the lattice parameters obtained in coexistence regions were sufficient to determine the directions of tie-lines. Figure 5.12 defines the angle α formed by a point in the coexistence region. For any given point, the spinel lattice parameter obtained in the two-phase fit can be used to find a corresponding point on the upper boundary of the spinel line with the same lattice parameter (i.e., this sample corresponds to the end of the tie-line). This point can then be used to define a theoretical tie-line forming angle α_c. Figure 5.12 (left) demonstrates this using the point (0.30, 0.66) which had two-phases present in the XRD and the spinel lattice parameter was 8.225 ± 0.003 Å. The contour plots from Fig. 5.7 were used in Fig. 5.12 to illustrate the calculation of α_c (in practice, the contour plots were not used, instead the values were obtained by extrapolating between the known data points). Ideally, $\alpha = \alpha_c$ for all points in the coexistence region. Figure 5.12 (right) shows the plot of α_c vs α obtained for samples heated in oxygen with regular cooling. Since calculating α_c involved extrapolating the spinel lattice parameters along the boundary of the single-phase region, the results were not perfect and the linear fit crosses slightly below the origin. Nonetheless, when a linear fit is forced through the origin, the result is $\alpha_c = 1.0006\alpha$ with a R-value of 0.9952 (R = 1 for a perfect linear fit). This result shows that with a sufficient number of samples in a coexistence region, the lattice parameters can be used to determine the tie-lines.

The position and shape of the upper boundary of the spinel region were determined using the lever rule. This method was discussed in Sect. 2.4, where a sample X made up of phases A and B was used to calculate the composition of phase B if the composition of phase A was known. In this chapter, point A was always at a composition

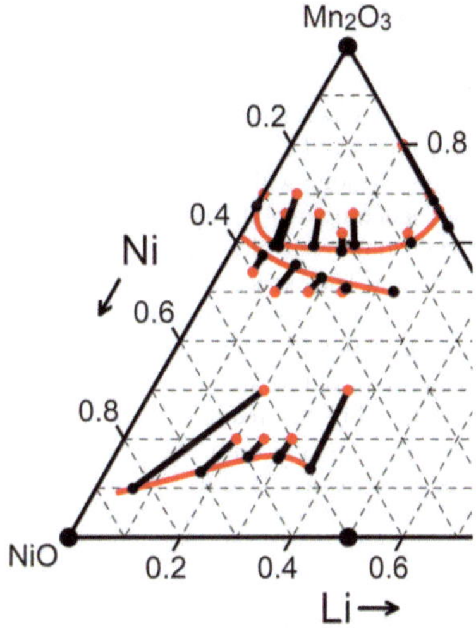

Fig. 5.13 The result of using the lever rule in the coexistence regions for samples prepared in oxygen by regular cooling. *Red points* represent the compositions of the samples in the two-phase region while the *black points* are the results of the lever rule calculations

from which tie-lines fanned out, e.g., Mn_2O_3 here, and the generated points, B, lie on the boundary of a single-phase region, e.g., upper spinel boundary. Best results were obtained by using points in the coexistence regions that were relatively close to the boundary being identified, thereby limiting the uncertainty resulting from the calculation. No corrections were made for microabsorption effects. Brindley suggested that this is justified for samples if $\mu D < 0.01$ for each phase present, where μ is the linear absorption coefficient and D is the diameter of the corresponding particles [82]. Only a few samples were examined by scanning electron microscopy (SEM) (shown in Chap. 6) but typical particle sizes were found to be in the 50–150 nm range. For $LiNi_{0.5}Mn_{1.5}O_4$, the absorption coefficient is about 570 cm^{-1} such that $\mu D = 0.0086$ for 150 nm particles which explains why using the lever rule without the Brindley correction worked well.

Figure 5.13 (left) shows the boundaries obtained by the lever rule for samples heated in oxygen with regular cooling. The boundaries were in excellent agreement with the visual identification of phases shown in Fig. 5.7a. The same method was used to generate the boundaries for the quenched samples as well as those heated in air and again the agreement with visually determined phases was good. The upper boundary of the spinel phase is shown in Figs. 5.2, 5.3, and 5.7.

5.5 Spinel–Ordered Rocksalt Coexistence Region

Figure 5.14 shows XRD scans in the coexistence region between the spinel and ordered rocksalt regions. The spinel peaks clearly diminished and were replaced by the ordered rocksalt peaks consistent with a two-phase regime. The fits shown in Fig. 5.14 show good agreement with the data and no peaks are unaccounted for.

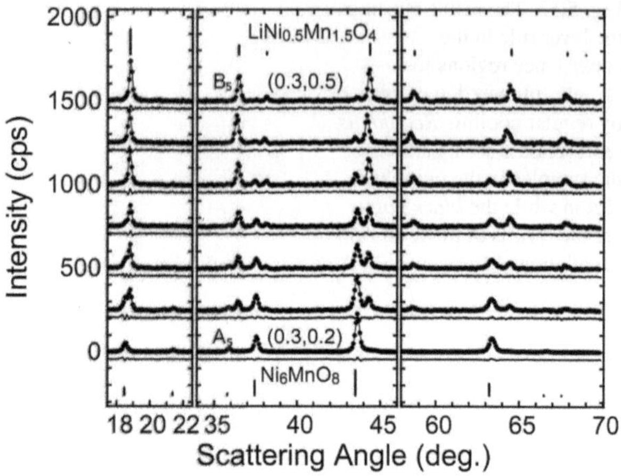

Fig. 5.14 X-ray diffraction (XRD) scans of samples in the coexistence region between the ordered rocksalt structures and the spinel structures along with the corresponding two-phase fits and the difference plots below each scan. The expected peaks for $LiNi_{0.5}Mn_{1.5}O_4$ are from JCPDS #80-2162. Samples A_5 and B_5 are labeled in Fig. 5.4 and the other samples are evenly spaced between A_5 and B_5

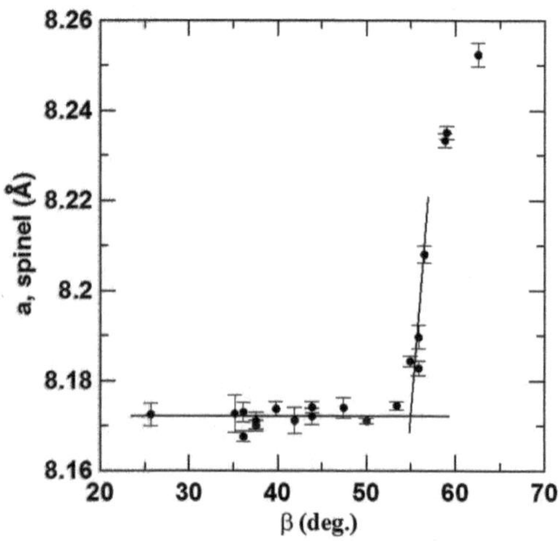

Fig. 5.15 The spinel lattice parameter obtained in the coexistence region between the spinel and ordered rocksalt regions as a function of β

Figure 5.15 shows the spinel lattice parameter obtained in the coexistence region by regular cooling in oxygen as a function of the angle β defined in Fig. 5.4. Up to $\beta = 55°$, the spinel lattice parameter remains constant and the 13 scans in this region have an average lattice parameter of 8.1722 ± 0.0005 Å with a standard deviation of 0.0019 Å which corresponds to 0.02 % of the lattice parameter. The average lattice parameter is therefore consistent with tie-lines fanning out from a single point. Furthermore, the value of the average lattice parameter is very close to 8.173 ± 0.001 Å, the lattice parameter obtained for $LiNi_{0.5}Mn_{1.5}O_4$, as shown in the contour plot of Fig. 5.7 (a). The tie-lines were therefore drawn fanning out from this point up to a maximum angle of $\beta = 55°$.

By contrast, in air, $LiNi_{0.5}Mn_{1.5}O_4$ is not single-phase but shows a small amount of a second phase as discussed in Sect. 1.4.1. The tie-line drawn at the composition of $LiNi_{0.5}Mn_{1.5}O_4$ in Fig. 5.3 shows that the spinel structure is coexisting

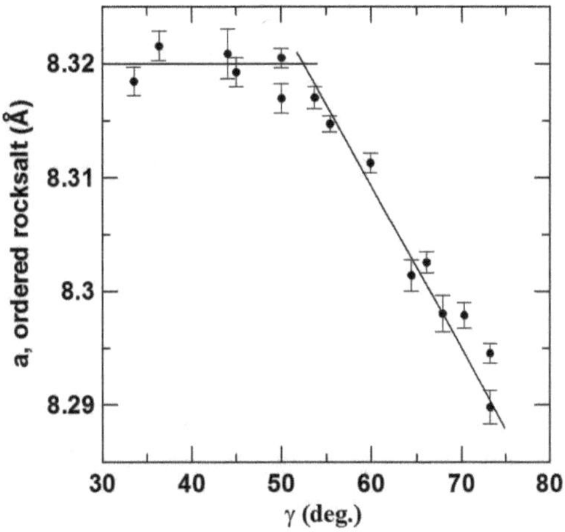

Fig. 5.16 The ordered rocksalt lattice parameter obtained in the coexistence region between the spinel and ordered rocksalt regions as a function of γ

with an ordered rocksalt structure quite close in composition to the one analyzed by Rietveld refinement in Sect. 5.3. This phase is previously been referred to as $Li_xNi_{1-x}O$ rocksalt [34]. Recognizing that the end of the tie-line is higher up in the triangle than previously suspected has consequences when one attempts to work out the stoichiometry of the spinel sample based on the amount of oxygen lost during synthesis.

The lever rule, described in Sect. 2.4, was used with $LiNi_{0.5}Mn_{1.5}O_4$ as the pivot point (point A in Fig. 5.13, right) in order to identify the upper boundary of the ordered rocksalt region. Figure 5.13 (left) shows the points used to define this boundary. Figure 5.10 shows that once again the boundaries obtained with the lever rule agree with the phases determined by visual inspection of the XRD scans. Although the point (0.45, 0.3) lies in the three-phase region of the regular-cooled triangle, it is close enough to the edge of the region that it appears two-phase and provides a useful point to obtain the correct curvature of the boundary of the rocksalt solid solution region.

The position of the boundary at $\beta = 55°$ is at (0.05, 0.09), the point used in Fig. 5.4 to define the angle γ. Figure 5.16 shows the ordered rocksalt lattice parameter as a function of γ. Here, the lattice parameter is constant with a value of 8.320 ± 0.001 Å up to a maximum angle of approximately 54°. This is close to the value of 8.328 ± 0.001 Å obtained for the sample at (0.05, 0.10), but this still represents the greatest discrepancy seen along tie-lines in this study. Should this region become of greater interest, it may be necessary to explore this boundary more closely. Nonetheless, the data suggest that the tie-lines fan out from a point near (0.05, 0.09) up to $\gamma = 54$ or 55° beyond which the samples were in the region where tie-lines fan out from $LiNi_{0.5}Mn_{1.5}O_4$ (i.e., $\beta < 55°$). All tie-lines in the region between the spinel and ordered rocksalt phases were therefore identified and included in Fig. 5.2.

The lever rule was then used with the point (0.05, 0.09) as the pivot in order to determine the lower spinel boundary to the left of $LiNi_{0.5}Mn_{1.5}O_4$. This was only done for samples heated in oxygen, and in both of these phase diagrams, this boundary matched up very well with the phases determined by visually examining the XRD scans, as well as joining nicely with the boundary determined in the region to the right of $LiNi_{0.5}Mn_{1.5}O_4$. In air, insufficient data were collected to use the lever rule in this region and so the lower spinel boundary in Fig. 5.3 is left incomplete.

5.6 Conclusions Regarding Spinel and Rocksalt Li–Mn–Ni Oxides

The entire spinel and rocksalt solid-solution regions of the Li–Mn–Ni oxide pseudo-ternary system were determined at 800 °C when regular cooled in air and when quenched or regular cooled in oxygen. All samples discussed here either contained one or two phases; and no evidence for nonequilibrium behaviour was seen. Two-phase fits in the coexistence regions were used to determine lattice parameters to allow the drawing of tie-lines. The lever rule was used to determine the boundaries and showed excellent agreement with visually identified phases. Milligram-scale combinatorial samples can therefore be used to obtain a high degree of precision in the mapping of pseudo-ternary phase diagrams.

The spinel samples near $LiNi_{0.5}Mn_{1.5}O_4$ quenched in oxygen showed oxygen vacancies. It is likely that samples quenched in air with a lower oxygen partial pressure also have vacancies. Since oxygen vacancy formation occurs simultaneously as an oxygen loss resulting in an upwards motion of the lower spinel boundary, it is important to recognize that both these effects are present in samples synthesized in air and are difficult to differentiate. However, it has not yet been determined whether or not bulk samples also show these deficiencies; it is possible that oxygen loss is significantly more prominent in the combinatorial samples due to the large surface area to volume ratios.

Comparing Figs. 5.2 and 5.3 shows the motion of single-phase boundaries with synthesis conditions. Generally, the single-phase regions are much larger than previously known and increase in size with temperature (i.e., quenched) and they are also larger when prepared in air. The ordered rocksalt boundary is higher in air than it is in oxygen. In air, the lower oxygen partial pressure favors structures with lower metal oxidation numbers (i.e., lower oxygen content) such that the boundary of the rock-salt and layered regions is expected to move upward, particularly when quenched. The resulting lattice parameter contour plots show that the spinel samples made in oxygen have oxygen site vacancies which are relieved during regular cooling.

The structure of the cubic rocksalt materials is more complex than previously suspected. The refinement of the XRD patterns showed ordering of lithium, manganese, and metal site vacancies on the cubic lattice, and the extent of this ordering increased during regular cooling.

Chapter 6
Combinatorial Studies of Compositions Containing Layered Phases in the Li–Mn–Ni–O System

6.1 Experimental Design

This chapter deals with the remainder of the Li–Mn–Ni–O phase diagram obtained with combinatorial samples. The experimental design is the same as that described in Chap. 5. The objective here is to show supporting evidence for the rest of the phase diagrams shown in Fig. 5.2.

A few combinatorial samples were imaged by scanning electron microscope (SEM) as described in Sect. 2.8. Bulk samples with key compositions were also synthesized using the "one-pot" method outlined in Sect. 2.1.2. The helium pycnometer was used to obtain accurate density measurements of an ordered rocksalt sample to calculate the concentration of vacancies present in the structure as described in Sect. 2.11.

6.2 Single-Phase Layered Region

Once again, one challenge with the combinatorial synthesis of samples in the layered region of the triangle is that the samples lose lithium during heating. Figure 6.1a shows the phase diagram with points resulting from using atomic absorption to determine the composition of the samples after heating. All points included here are within error bars from the as-dispensed composition. Other samples were synthesized by the same method and characterized by X-ray diffraction (XRD) in order to determine the phases present. The boundary line from $LiNiO_2$ to Li_2MnO_3 is drawn as a straight line as this seems to agree with the data available but more samples are required to confirm this. The upper boundary of the layered region was determined as described in Sect. 6.8 and was in good agreement with the samples shown in Fig. 6.1a.

Data in this chapter are reprinted from Ref. [12] with permission from the American Chemical Society.

E. McCalla, *Consequences of Combinatorial Studies of Positive Electrodes for Li-ion Batteries,* Springer Theses, DOI 10.1007/978-3-319-05849-8_6,

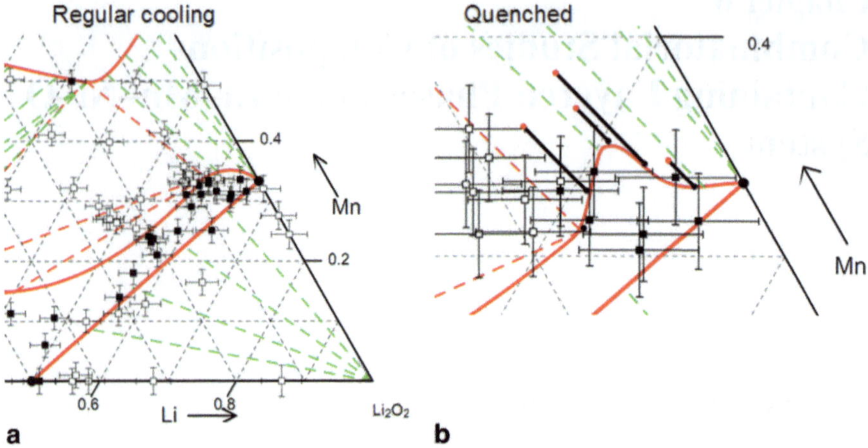

Fig. 6.1 **a** Metal compositions obtained by atomic absorption for samples prepared by regular cooling in oxygen. The *open symbols* indicate multiple phases in the XRD patterns while the *closed symbols* indicate single-phase scans. **b** A portion of the phase diagram obtained in oxygen by quenching. The compositions are assumed to be identical to those determined by atomic absorption for samples prepared by regular cooling. The *red points* are two-phase samples connected to the point obtained by the lever rule to generate the edge of the layered region

Figure 6.2 shows contour plots for both the a and c lattice parameters when the structures in the layered region are fit using peak indexing according to a lithium-rich layered material, $Li_{1.15}(Mn_xNi_{1-x})_{0.85}O_2$ (JCPDS #52-0457) [83]. In the region near the Li–Mn line, the addition of lithium results in a significant reduction in the c axis but very little change in the a lattice parameter. Since this is part of the lithium-rich region, this increase in lithium content is likely to result in an increase in the amount of lithium in the transition metal layer such that the hexagonal layers stack tighter while the spacing within the layers is mainly unaffected. The a and c values obtained for $Li_xNi_{2-x}O_2$ near $x = 1$ are in good agreement with Li et al. [36].

6.3 Two-Phase Layered–Spinel Region

Figures 6.3 and 6.4 (right) show XRD scans near the Li–Mn edge of the Gibbs triangle, at compositions indicated in Fig. 6.5 where the spinel and layered phases coexist. The H_6 sample at (0.6, 0.35) is single-phase consistent with the boundary that will be established in Sect. 6.8. All other samples show both spinel and layered peaks which are all well described by the two-phase fits. The scattering angle range 43–46°, emphasized in Fig. 6.4 (right) clearly shows the relative amounts of the two phases, and this region will be of use in illustrating the phases present throughout this study.

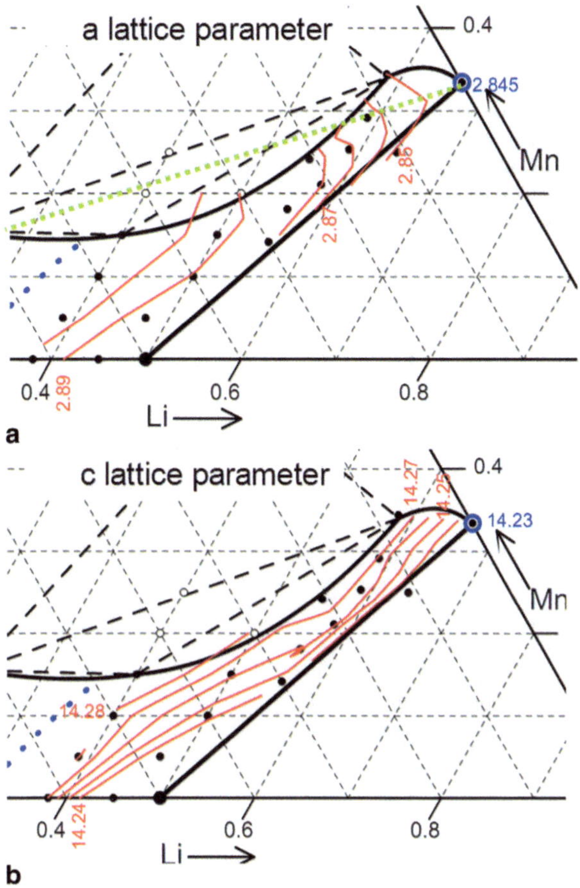

Fig. 6.2 Contour plots of the *a* and *c* lattice parameters obtained by fitting all regular cooled layered structures made in oxygen as hexagonal structures. The *green dotted line* in **a** is the constant oxidation number line for layered structures with Ni^{2+} and Mn^{4+}. *Filled points* represent single-phase samples while *open symbols* represent multiple-phase samples

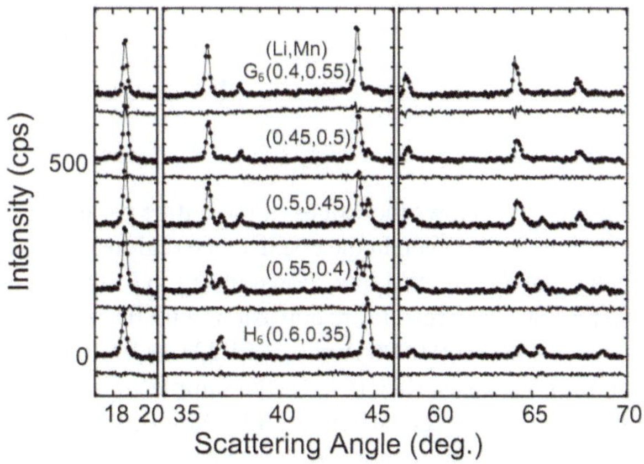

Fig. 6.3 XRD scans of quenched samples in the coexistence region showing both layered and spinel structures near the Li–Mn line. The compositions of samples H_6 and G_6 are indicated in Fig. 6.5

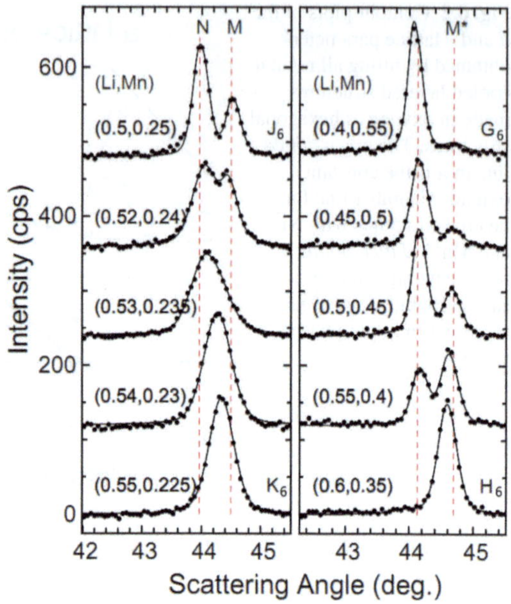

Fig. 6.4 XRD scans of quenched samples in the layered–layered (*left*) and layered–spinel (*right*) coexistence regions. J_6, K_6, G_6, and H_6 are indicated in Fig. 6.5. M^* represents Li_2MnO_3 and the *unlabeled dashed line* is a guide for the eye near the spinel peak. *N* and *M* correspond to two corners of the three-phase regions

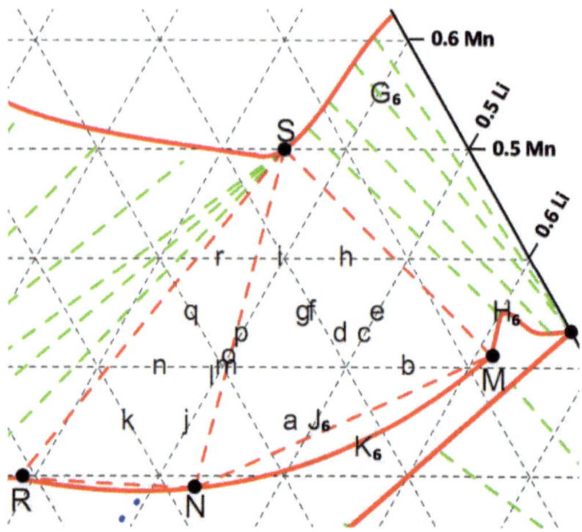

Fig. 6.5 A partial phase diagram indicating compositions used in Figs. 6.4, 6.7, 6.10, and 6.11

Figure 6.6 shows the values of the lattice parameters obtained for quenched samples heated in oxygen in the coexistence region as a function of the distance between each point and the SM line shown in Fig. 6.5. These graphs were used to determine the tie-lines. The results are consistent with tie-lines first fanning out from Li_2MnO_3 ($a = 2.8438 \pm 0.0006$ Å, $c = 14.231 \pm 0.007$ Å for a quenched combinatorial sample). The point (0.5, 0.45) is the sample with the most nickel that was found with lattice parameters consistent with Li_2MnO_3, thus this was used to establish the end of the fanning out region. Further from the Li–Mn edge, the tie-lines move in such

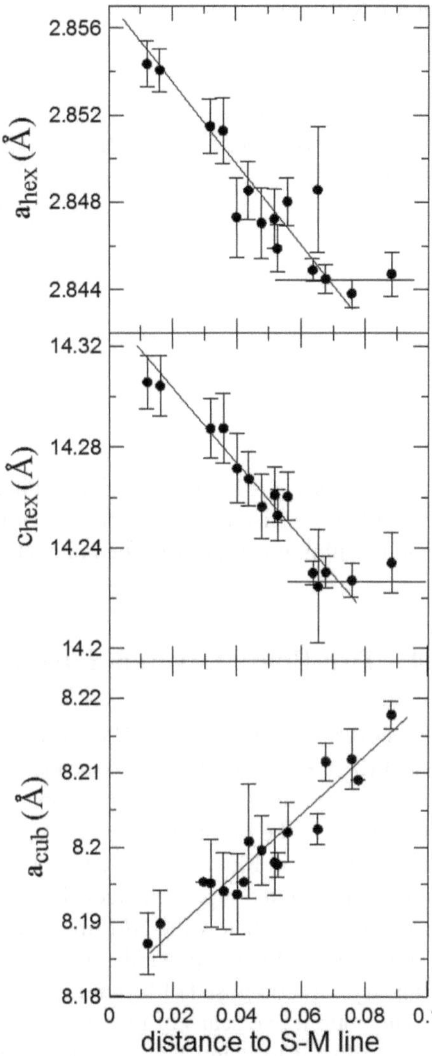

Fig. 6.6 Lattice parameters obtained in the layered–spinel coexistence region as a function of the distance to the SM line. The spinel lattice parameter is a_{cub} while the layered lattice parameters are a_{hex} and c_{hex}. The *lines* are guides for the eye

a way that both endpoints change, resulting in all three lattice parameters changing nearly linearly in this region in Fig. 6.6. Since the tie-line through (0.5, 0.45) is nearly parallel to the SM line, it is reasonable to assume that the tie-lines continue to run parallel to the SM line and this is how they were drawn in Fig. 5.2.

The lever rule was used, as described previously in Sect. 2.4, for the quenched samples in order to help determine the single-phase boundary of the layered region between the M-layered phase and Li_2MnO_3. Figure 6.1b shows the points generated by this method as well as the positions of samples as determined by atomic absorption. Clearly, these observations are nearing the precision limits of the elemental analysis,

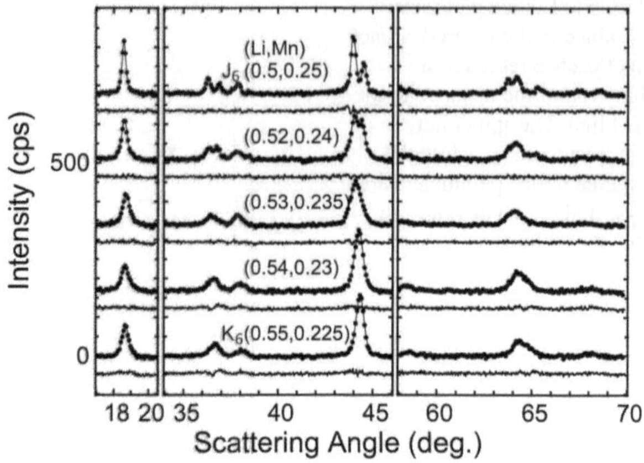

Fig. 6.7 XRD scans in the two-phase region below the NM line obtained for quenched samples. Compositions of samples J_6 and K_6 are indicated on Fig. 6.5

but nonetheless the boundary generated in Sect. 6.8 to fit the points obtained by the lever rule is consistent with all but one of these data points and this one anomalous point is within uncertainty of the boundary.

6.4 Two-Phase Layered–Layered Region

Figures 6.4 (left) and 6.7 show XRD scans of quenched samples lying between points J_6 and K_6 in Fig. 6.5. The top two scans clearly show multiple phases present based on the region near 44°. The top sample is at the composition of $LiNi_{0.5}Mn_{0.5}O_2$ and appears to be on the tie-line between the M and N points, consistent with the phase diagram in Fig. 6.5. The reader is reminded that these are measurements for samples heated in oxygen and quenched. This left only one scan, (0.52, 0.24), in the two-phase region below the NM line and this was insufficient to generate tie-lines. It should be kept in mind that the region between the NM line and the upper layered boundary was composed of layered–layered composites.

The bottom two scans of Fig. 6.4 (left) both appear to be single-phase. The one that is difficult to identify is the sample at (0.53, 0.235) where severe broadening was seen but two peaks cannot yet be distinguished. This suggests that two phases were present in this sample, and this was supported by the fact that the extra peak seen in the scan at (0.52, 0.24) appeared on the side where there is an asymmetry to the broad peak in (0.53, 0.235). The boundary was therefore placed at (0.535, 0.2325), halfway between the single and two-phase samples. Despite this analysis, the exact position of this boundary should still be considered in question to within about 0.01 in lithium content.

In samples cooled at the regular rate the XRD patterns look similar to that seen when quenched except for the one synthesized at point J_6 which shows evidence of the R phase as well as the N and M phases. This behavior will be discussed further in Sect. 6.7.

6.5 The R, N, S, and M phases

Figure 6.5 shows two three-phase regions side-by-side between the spinel and layered single-phase regions when samples were heated in oxygen and quenched. The two regions were the NSM and NSR triangles. R was an ordered rocksalt structure, S was a spinel structure, and N and M were both layered structures (labels were chosen to represent nickel-rich or manganese-rich layered structures to distinguish them). The large array of samples shown in Fig. 5.1b made it possible to identify samples near each corner by matching lattice parameters to those obtained with the three-phase fits. Figures 6.8 and 6.9 (left) show XRD scans of quenched samples near the R, S, N, and M corners. Fitting these patterns allowed the determination of the lattice parameters included in Table 6.1 and their coordinates in the Gibbs triangle are listed in Table 6.2. In the scattering angle range 43–46°, each of the four single-phase samples have a single peak without $K_{\alpha_{1,2}}$ peak splitting. As such, this region is ideal to illustrate the phases present in the three-phase regions.

Bulk quenched samples synthesized at each of the R, N, and M corners using the one-pot synthesis (Sect. 2.1.2) were scanned with the JD-2000 diffractometer. Table 6.3 shows the results of Rietveld refinement. The R phase only has 10 % vacancies such that the average Mn oxidation state is 3.6+, assuming that nickel is in the 2+ state. The density of this R phase sample was found to be 5.4447 ± 0.0090 g/mL using a helium pycnometer and this value corresponds to 8.8 ± 2.0 % vacancies on the 4b sites, in good agreement with the XRD results. In the previous chapter, an ordered rocksalt structure synthesized in oxygen and cooled at the regular rate was found to have approximately 30 % vacancies on the 4b sites and Mn in the 4+ state. These results suggest that the structure of this phase changed during slow cooling and further study is required to understand this fully.

The N-layered phase had a hexagonal structure with significant disorder in both the lithium and transition metal layers such that there was a 30 % nickel occupation on the lithium layers. By contrast, the M-layered phase was far more ordered with very little nickel on the lithium layers. XRD scans of this structure also show superlattice peaks in the range 20–35° consistent with ordering on $\sqrt{3} \times \sqrt{3}$ lattices in the transition metal (TM) layers. Although the compositions of the TM layers were not in the 1:2 ratio needed for this ordering, it was quite close. As such, if one $\sqrt{3} \times \sqrt{3}$ lattice was randomly occupied with 0.708Li and 0.292Ni while the other two contained 1.866Mn and 0.134Ni, the resulting scattering contrast would be sufficient to give rise to the weak ordering peaks seen here, consistent with the reasoning of Lu et al. [45]. The ordering peaks should, therefore, not necessarily be interpreted as the presence of Li_2MnO_3 and the sample may yet be single-phase as suggested here.

Although the ordering seen in the M sample is expected to be coupled with monoclinic distortions in the hexagonal lattice, the fact that a high-quality fit was obtained for a hexagonal structure shows that the extent of monoclinic distortion was small and it was completely masked by the peak broadening due to the diffractometer. The fact that the oxygen occupancy converges to a value above 100 % implies that there are metal site vacancies in the M-layered structure. Both of these issues will be resolved in Chap. 8.

Table 6.1 The lattice parameters obtained for the single-phase corners of the three-phase regions as well as the average values obtained in the three-phase regions and the standard deviations (σ). All values are in Å. The uncertainty for the average lattice parameters is calculated as $\sigma/\sqrt{N}$ where N is the number of samples

	Rocksalt		Spinel		N-layered		M-layered	
	a	σ	a	σ	a c	σ	a c	σ
Quench:								
single phase	8.284 (2)		8.187 (1)		2.909 (1) 14.32 (1)		2.8543 (6) 14.291 (6)	
SNM (26 fits)			8.1867 (9)	0.0044	2.9025 (6) 14.347 (2)	0.0033 0.010	2.8553 (2) 14.303 (2)	0.00093 0.002
RSN (12 fits)	8.284 (3)	0.0097	8.182 (2)	0.0070	2.902 (2) 14.322 (5)	0.0053 0.017		
Slow cool:								
single phase	8.295 (2)		8.173 (1)		2.907 (1) 14.30 (1)		2.8516 (8) 14.277 (8)	
MRS (6 fits)	8.297 (2)	0.0052	8.169 (1)	0.0018			2.8502 (5) 14.280 (6)	0.0012 0.013
MRN (12 fits)	8.294 (2)	0.0080			2.9013 (4) 14.312 (4)	0.00014 0.012	2.854 (2) 14.25 (1)	0.0072 0.032

Table 6.2 The coordinates of the four corners of the three-phase regions. All points are (Li, Mn)

	Rocksalt	Spinel	N-layered	M-layered
Quench:				
single-phase	(0.3, 0.2)	(0.36, 0.5)	(0.4, 0.17)	(0.61, 0.31)
Lever rule	(0.29, 0.20)	(0.36, 0.5)	(0.43, 0.19)	(0.61, 0.31)
Slow cool:				
single-phase	(0.22, 0.16)[a]	(0.33, 0.5)	(0.4, 0.15)	(0.58, 0.34)

[a] This point lies midway between the 8.29 and 8.30 Å contour lines on the boundary of the ordered rocksalt region in Fig. 5.10

Table 6.3 Rietveld results for bulk samples synthesized at the R, M, and N corners by heating in oxygen for 5 h and quenching to room temperature. Occupations and positions of each site are shown, as well as lattice parameters in Å

R (Fm3m)	N (R−3m)	M (R−3m)[a]
a = 8.2829 (4)	a = 2.916 (1) c = 14.324 (6)	a = 2.8574 (5) c = 14.289 (4)
4a: 4.0 Mn 4b: 3.272 Li 0.328 Ni 0.4 □	3a: 0.499 Li 1.134 Mn 1.367 Ni	3a: 0.708 Li 1.866 Mn 3a: 0.426 Ni
8c: 8 O 24d: 6.084 Li 2.176 Mn 15.74 Ni	3b: 2.105 Li 0.895 Ni	3b: 2.976 Li 0.024 Ni
24e: 24 O (0.2378,0,0)	6c: 5.588 O (0.2564,0,0)	6c: 6.253 O (0.2579,0,0)
R_B= 1.99 %	1.86 %	2.92 %

[a] This ignores the superlattice peaks which can be indexed to the C2/m space group.

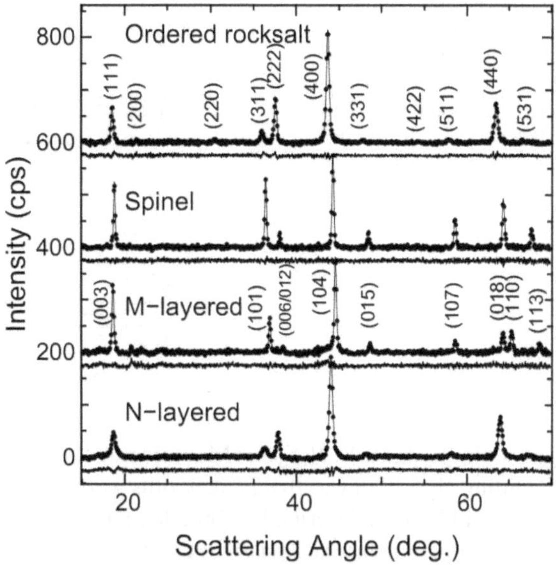

Fig. 6.8 XRD patterns of the four samples found at the corners of the three-phase regions for samples heated in oxygen and quenched. The peak indices are indicated based on JCPDS #89-4619 for the *ordered rocksalt* phase and #52-0457 for the *layered* phases. The *spinel* peaks have the same indices as the *ordered rocksalt;* though there are no (200), (331), and (422) reflections in the *spinel* structures and the (220) peak is weak. The *M-layered* structure is indexed to a hexagonal structure as discussed in the text. The *N-layered* peaks have the same indices, though the (018) and (110) peaks are not separated

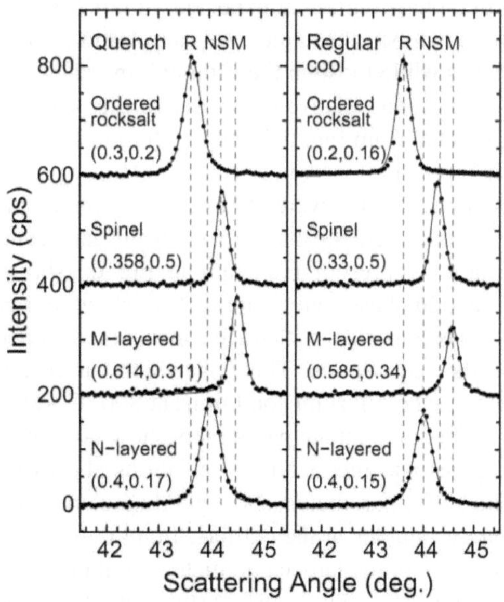

Fig. 6.9 XRD patterns generated at the four corners of the three-phase regions obtained by quenching (*left*) and regular cooling (*right*). The coordinates are (Li, Mn) where Li is the lithium metal fraction and Mn is the manganese metal fraction

6.6 Three-Phase Regions, Quenched

Figure 6.10 shows a number of XRD scans obtained within the three-phase regions of quenched samples. The patterns were complex with a number of overlapping peaks. The whole pattern fits were obtained as linear combinations of the four fits generated

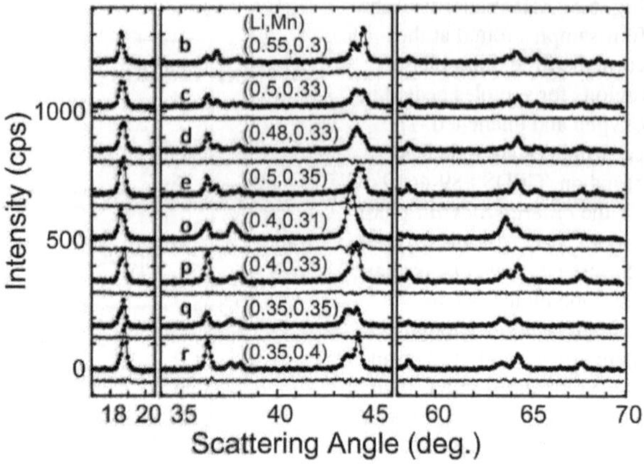

Fig. 6.10 XRD scans of samples in the three-phase regions obtained by quenching. The fits are included as well as difference plots below each scan. The labels (*b, c, d* . . .) match those used in Fig. 6.5

for the single-phase R, N, S, and M samples in the previous section. The key feature of the fitting algorithm was that peak widths were never adjusted such that overlapping peaks could not be described by broadening the peaks of a particular phase. Instead, the lattice parameters and phase intensities were allowed to be adjusted. The lattice constants of the three coexisting phases were virtually invariant for all the fits in the three-phase regions as expected. Figure 6.11 shows the 43–46° scattering angle range for a number of samples. All samples in the left column show primarily N, S, and M peaks while those in the right column show that N, S, and R peaks dominate. In all samples, the relative intensities of the peaks were qualitatively consistent with that expected from Fig. 6.5. Although the intensities of the peaks changed significantly, their positions were nearly constant throughout Fig. 6.11, consistent with three-phase regions. The fits were obtained using three phases only so that sections of the XRD patterns where the fits were below the data revealed where trace amounts of the fourth phase was present. For example, Fig. 6.5 shows that point f should be made up of N, S, and M phases only, but Fig. 6.11 shows a small peak corresponding to the R phase. This is consistent with trace conversion occurring during cooling as will be discussed in the next section. This conversion can be attributed to imperfect quenching such that the equilibrium conditions at 800 °C were not maintained during cooling.

The fitted lattice parameters show no trends with composition in the three-phase regions and the standard deviations shown in Table 6.1 are small. The values of the lattice parameters show good agreement with the values obtained for the single-phase samples at the corners. In all cases, the standard deviations are on the order of 0.1 % of the lattice parameter and therefore small enough to imply that the lattice parameters remain constant (this condition was determined in the previous chapter). This supports the claim that the NSM and NSR triangles are in fact three-phase regions.

Fitting the three-phase scans also allowed for the use of the lever rule to determine the tie-lines lying at the outer edges of the three-phase regions. The method was similar to that used in two-phase regions wherein phase fractions were calculated from fitted peak areas in the coexistence and single-phase regions. Precise values

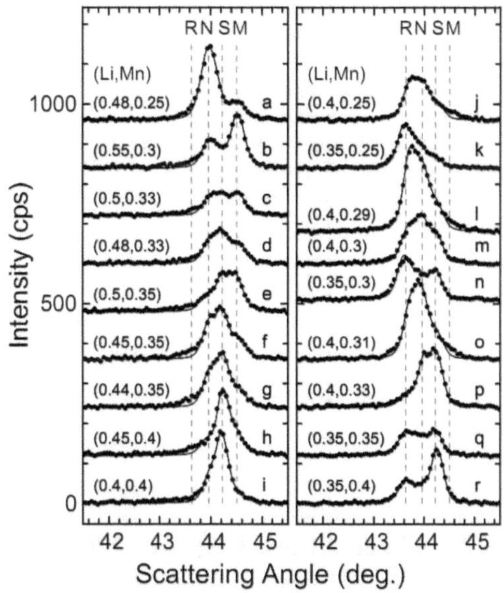

Fig. 6.11 Partial XRD scans of samples in the three-phase regions obtained by quenching. *Left*: samples in the NSM triangle. *Right*: samples in the NSR triangle. The labels (*a*, *b*, *c* ...) match those used in Fig. 6.5

for the peak areas of the four corners were obtained from the fits shown in Fig. 6.8. The sequence used to generate the tie-lines at the outer edges of the three-phase regions began by using the S corner as the pivot point for the lever rule in order to define the RN and NM lines. Figure 6.12 shows this use of the lever rule with black lines joining a three-phase sample in red to the calculated point on the RN or NM line in black. The intersection of these two lines gave the N point which was then used as the pivot in order to determine the SR and SM lines. These lines allowed the identification of the R and M corners. Finally, the R and M corners were used as pivots to define the MS line. The intersections of the MS, RS, and NS lines were then used to verify the original choice of corner S. Figure 6.12 shows the results of using the lever rule to generate the boundaries and identify the locations of the corners. Many of the three-phase points could be used to generate all three boundaries, however some were found to be either too close or too far from the boundary to give accurate results. Despite this limitation, there are sufficient points to tightly constrain each boundary and therefore yield precise coordinates for the four corners. Table 6.2 shows that the coordinates of the four corners agree well with those obtained by searching for single-phase samples with lattice parameters matching those obtained in the three-phase regions.

These results show that the phase compositions observed in the regions are consistent with the three-phase regions shown in Figs. 5.2a, 6.5, and 6.12. However, imperfect quenching did result in trace contamination with a fourth phase being present in some samples.

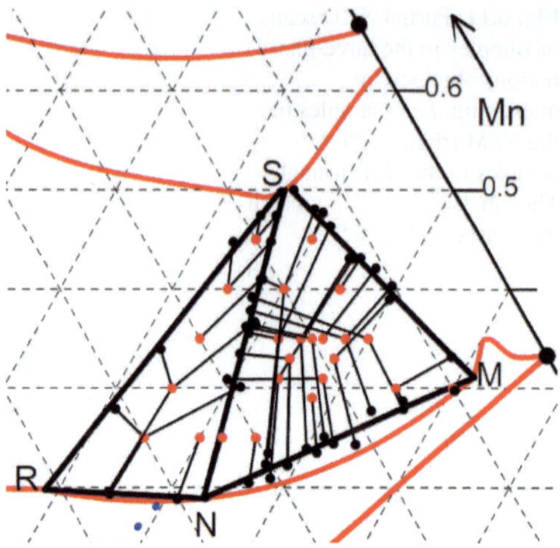

Fig. 6.12 A partial phase diagram focusing on the three-phase regions for quenched samples. Samples represented by the *red points* show three-phases and they are connected to *points on the tie-lines* as calculated with the lever rule

6.7 Three-Phase Regions, Slow-Cooled

The transformations during slow cooling are complex, and all samples in the three-phase regions change dramatically. Figure 6.13 shows partial XRD scans obtained by quenching, regular cooling (about 8 °C/min) and slow cooling (1 °C/min). The compositions of the samples in the left panel are equally spaced along a line joining A_6 and B_6 in Fig. 6.14 while the right panel represents samples equally spaced between points C_6 and D_6. All quenched samples were consistent with Fig. 6.12 (i.e., they have the expected amounts of R, N, S, and M phases). For all samples along the A_6–B_6 and C_6–D_6 lines, the fraction of N-layered phase diminished as cooling time increased. Although a small amount of N remained in some samples when slow-cooled, it appears that all of these samples were tending toward having R, S, and M phases only; consistent with XRD scans found in the literature for regular cooled samples [13].

Figure 6.9 (right) shows the XRD scans of regular-cooled single-phase samples identified near the four corners, which have moved relative to their positions in the quenched samples. Table 6.1 shows the fitted lattice parameters for these samples and Table 6.2 shows their locations within the Gibbs triangle. Figure 6.15 shows XRD scans and fits of a few samples obtained by regular cooling which appeared to have finished conversion (i.e., with only three peaks present in the region 43–46°). The fits clearly are poorer quality than those obtained for quenched samples, consistent with the fact that there were still significant amounts of the fourth phase present, and the fits included three phases only. Figure 6.16 shows the 43–46° region of XRD patterns from a variety of samples located as shown in Fig. 6.17. The data suggests that at some temperature during cooling the equilibrium diagram is made up of the following three-phase regions: MRS and MRN. It is of interest that conversion occurred more rapidly in the region near the SM line where the change primarily involved the replacement of the N phase with the R structure.

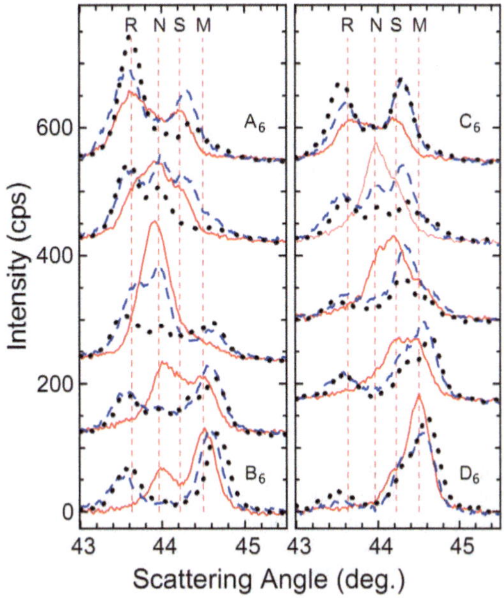

Fig. 6.13 XRD scans obtained in oxygen by quenching (*red-solid*), regular cooling (*blue-dashed*), and slow cooling (*black dotted*). The *vertical red dashed lines* indicate the positions of the peaks obtained by quenching

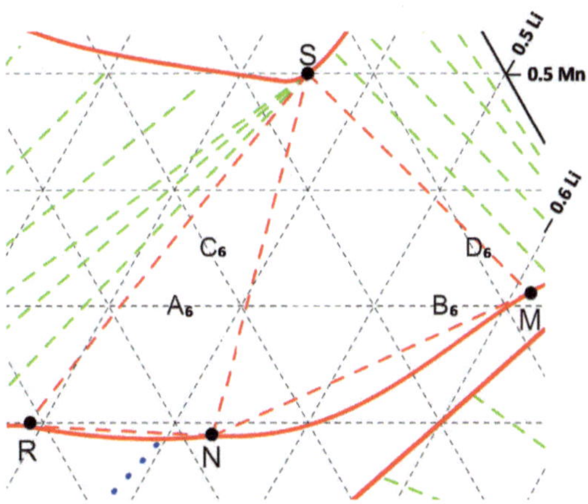

Fig. 6.14 A partial phase diagram with useful composition labels used throughout this chapter

The complex phase transformations occurring in the three-phase regions were seen once before in the literature by Hinuma et al. [55] as shown in Sect. 1.4.3. The sample synthesized at the composition of $LiNi_{0.5}Mn_{0.5}O_2$ by ion-exchange, which the authors suggested results in a slight lithium deficiency. Extra peaks appeared after annealing at 600 °C. Zooming in on the scattering angle region of 43–46 ° in the XRD scans (Fig. 8a in Ref. [55]) shows a curve qualitatively similar to scan d from Fig. 6.16 obtained here. This shows that the annealing step resulted in a phase diagram where the sample was positioned within the MRN triangle as obtained here by slow cooling in oxygen. Assuming that their sample was indeed lithium deficient, the results of Hinuma et al. are consistent with the phase diagrams presented here.

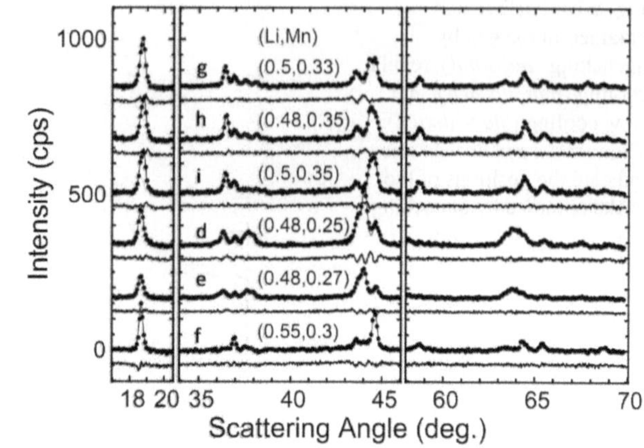

Fig. 6.15 XRD scans of samples in three-phase regions obtained by regular cooling with fits and difference plots below each scan. The labels (*d, e, f* . . .) match those used in Fig. 6.17

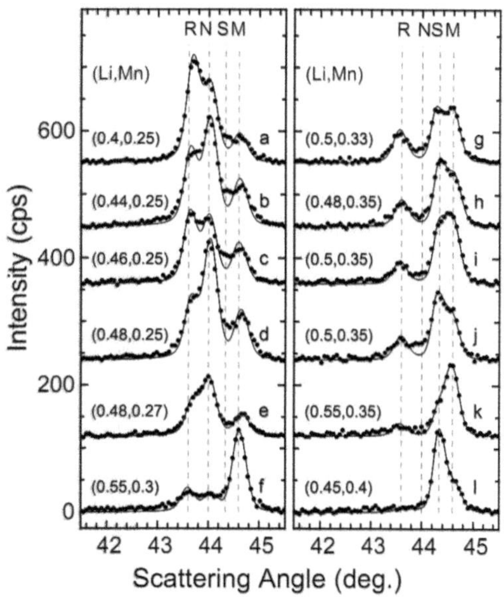

Fig. 6.16 Partial XRD patterns obtained in the three-phase regions by regular cooling. The labels (*a, b, c* . . .) match those used in Fig. 6.17

Figure 6.18 shows SEM images for three samples of interest as well as the corresponding XRD scans near 44°. Fig. 6.18c shows a sample at composition (0.5, 0.2) shows small particles consistent with the peak broadening seen in the XRD patterns. The sample imaged in Fig. 6.18a showed three phases in the XRD pattern consistent with the R, N, and M phases discussed here. The SEM image of this also showed similar particles to those seen in Fig. 6.18c, but a second image Fig. 6.18b of the (0.5, 0.25) sample showed that there are also a few large particles showing 90° crystal faces. These large particles may have been the ordered rocksalt, since being few in number and quite large would account for the small relatively sharp rocksalt peak in the XRD pattern. Such large ordered rocksalt particles would also be consistent with the fact that the ordered rocksalt phase grew rapidly during slow cooling of samples

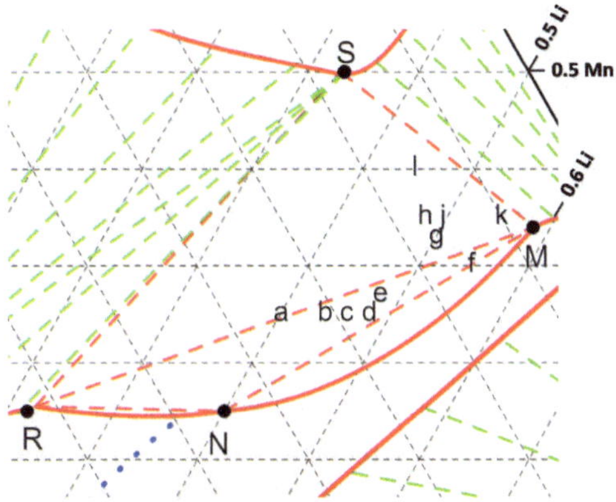

Fig. 6.17 A partial phase diagram showing compositions used in Figs. 6.15 and 6.16

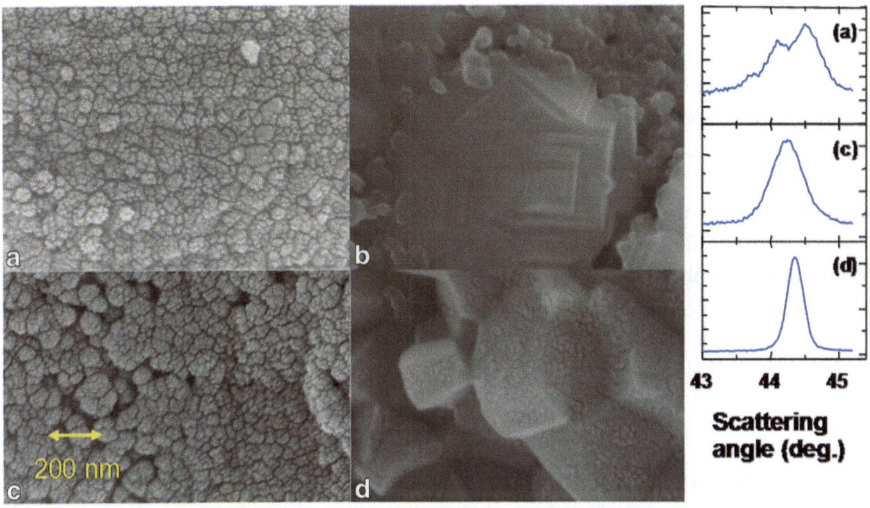

Fig. 6.18 *Left*: SEM images taken by Ramesh Shunmugasundaram of three combinatorial samples made in oxygen with regular cooling. **a** and **b** are two images of a sample with (Li, Mn) = (0.5, 0.25). **c** (0.5, 0.20) near the layered boundary. **d** is a sample dispensed at (0.5, 0.05). *Right*: XRD scans of the three samples in the region near 44°

in the three-phase regions. Although more study is required here, the SEM images do appear to support the observations made based on the XRD patterns.

6.8 The Upper Boundary of the Layered Region

The upper boundary of the layered region was not easy to determine due to the three-phase regions. The lever rule in the three-phase regions only yielded three points on the layered boundary: R, M, and N and this could only be done for quenched samples since the slow-cooled samples never reached equilibrium. For the regular cooled samples, the boundary was made using a polynomial function through five points: the R, M, and N corners, Li_2MnO_3 (which was on the boundary as determined in Sect. 6.3), and the single-phase point found on the edge of the layered–layered coexistence region (Sect. 6.4). Since only five points were used, the precision of this line is limited, particularly in the region below the RN line, where no samples were synthesized in the small two-phase region that must exist below the three-phase NSR region. It should also be noted that in the three-phase regions equilibrium is not reached during regular cooling so this could be the case in the layered–spinel two-phase region as well, making the layered boundary obtained by slow cooling particularly difficult to identify. Nonetheless, the boundary shown in Figs. 5.2b and 6.1a is in good agreement with the data at hand.

Figure 6.1b shows that for quenched samples the region near Li_2MnO_3 shows a sharper "bump" than that seen for slow cooled materials. The points generated in Sect. 6.3 using the lever rule were used in addition to the five points mentioned here. The narrow "bump" seems strange in contrast to the rest of the boundaries which are smooth broad curves. Again this feature is based on a relatively small number of data points and the quenching required about a minute suggesting that the high temperature structure may not have been frozen in. As such, more data was required to conclude that this feature was in fact the equilibrium boundary at 800 °C. This region will, therefore, be examined in more detail in Chap. 8. The phase diagram shown in Fig. 5.2a is in agreement with all quenched samples synthesized with the combinatorial robot. It is also worth noting that the boundaries of the single-phase regions were obtained by fitting polynomials to a relatively small number of points. The curvature at some points may therefore not be perfect, but the boundaries are consistent with all experimental data.

6.9 Conclusions Regarding Combinatorial Studies of Li–Mn–Ni–O Materials

The phase diagrams obtained for quenched samples show that in both three-phase regions all samples contain some of the N and S phases, while during slow cooling they convert to structures containing at least some R and M phases. Though the thermodynamics of this are complex, a qualitative understanding can be achieved by considering the results of the Rietveld refinement. The R and M phases were collectively more ordered than the N and S phases (the N phase in particular showed considerable disorder in the hexagonal layers). This would suggest that the R and M phases have lower combined internal energies and lower combined entropies than the combination of the N and S phases. Since the structure of the three-phase regions at high temperature is driven by entropy, the N and S phases are present in all points in the regions. Upon slow cooling, a temperature is reached where internal energy becomes more important and there is still sufficient thermal energy for atomic transport such that the R and M phases begin to appear in all samples. Further study is required to have a better understanding of the thermodynamics and kinetics of these phase transformations.

The results obtained here along the lithium rich line between $LiNi_{0.5}Mn_{0.5}O_2$ and Li_2MnO_3 also require discussion. This composition line is not a solid solution over its entire length when synthesized in oxygen. This would suggest that the broadening seen in XRD patterns by Jo et al. [54] was due to multiple phases. Furthermore, the region near 43.5° in the XRD spectra published by Jo et al. shows that there may be trace amounts of ordered rocksalt present, which means that this sample obtained with a slow cooling rate contained three phases, consistent with the phase diagram produced here.

Furthermore, the layered boundary moved upward in the Gibbs triangle when heated to higher temperatures (i.e., quenched), especially near the N-layered corner. At 800 °C in oxygen, a layered material synthesized at the composition of $LiNi_{0.5}Mn_{0.5}O_2$ is not stable, but instead, the phase separates into two layered structures: the N and M phases. As the boundary moves up (either with temperature or in air) the endpoints of the tie-line on which the sample lies will approach each other and ultimately become a single-phase sample. Inversely, during cooling the boundary moves down such that a single-phase sample near the upper boundary of the layered region phase separates into two layered structures during slow cooling. As suggested from the work in the Li–Co–Mn–O system in Chap. 4, these conditions would give rise to layered–layered nano-composites. However, the endpoints of this coexistence are not Li_2MnO_3 and $LiNi_{0.5}Mn_{0.5}O_4$ as assumed by many authors [43]. The tie-lines obtained here never extend to Li_2MnO_3 and they include points to the left of $LiNi_{0.5}Mn_{0.5}O_2$ in the Gibbs triangle. Therefore, the so-called $LiNi_{0.5}Mn_{0.5}O_2$-Li_2MnO_3 nano-phase separation promoted by many authors needs to be adapted given the actual endpoints of the coexistence. The layered–layered region does not include the lithium-rich materials at all at higher temperatures as the single-phase boundary sweeps upward in the triangle though. This will be discussed in the next chapter. It is also relevant that the unit cell volumes of the M and N-layered phases differ by approximately 4.0 %; which is relatively large compared to that observed in Chap. 4 and would most likely result in fracturing of the lattice. This prediction is supported by the SEM images here where very small particles are seen in the samples with two layered structures and larger particles in the single-phase layered sample. Thus, the region where nano-scale domains would be found is further restricted to some portion of the layered–layered region. Clearly, further study is needed to truly understand the short-range ordering in such samples and any sample near the boundary of the layered region where XRD peak broadening is seen. In particular, the consequences of these nano-domains on the lithium layer must be evaluated given that one of the layered structures in the nano-composites must be quite close in composition to the N-layered structure which was found to contain a high fraction of nickel on the lithium layer. Clustering of nickel on the lithium layer during cooling would interfere significantly with lithium extraction during electrochemical cycling. This will be explored in Chap. 9 for materials near $LiNi_{0.5}Mn_{0.5}O_4$.

This chapter should be of particular benefit to researchers working on composite electrode materials in the Li–Mn–Ni–O system. The phase diagrams produced here are consistent with all published XRD patterns found in the literature for samples in the coexistence region between the spinel and layered regions. At this point in time, it is unclear what dominates the properties of composite electrodes and in particular how the respective phases affect each other during cycling. Having a phase diagram should now make it possible to perform systematic studies of spinel-layered composite electrodes to better understand how the properties of a composite relate to those of its component phases.

Chapter 7
Investigations of Bulk Li–Mn–Ni–O Samples to Confirm the Combinatorial Studies

7.1 Motivation

Figure 7.1 shows isothermal sections of the Li–Mn–Ni–O phase diagrams obtained for the milligram-scale combinatorial samples heated in oxygen, the supporting evidence for which was presented in the last two chapters. Points A_7 and B_7 refer to compositions studied here in order to confirm that the results of the combinatorial work are relevant to bulk samples, and to help identify how the phase diagrams change for bulk samples. The single-phase regions in the combinatorial studies were larger than previously suspected with an expanded layered region that encompassed areas both above and below the Li-rich layered line joining Li_2MnO_3 to $LiNi_{0.5}Mn_{0.5}O_2$. This has significant consequences both in terms of reinterpreting published data and obtaining a better overall understanding of the electrochemistry of these materials. Some of these consequences will be explored in this chapter by studying bulk samples synthesized at the compositions listed in Table 7.1.

A sample made at point A_7 in Fig. 7.1 should be primarily made up of the N and S phases when quenched and convert to R and M during slow cooling. Rhines referred to such a transformation as a ternary four-phase equilibrium [85], though it is of note that oxygen content is changing here such that it is a more complicated transformation than that considered by Rhines. Nonetheless, he predicted that such a point would be present between two three-phase regions transforming during cooling as seen in Chap. 6. Such a point transforming reversibly would be sufficient to demonstrate that the three-phase regions exist and transform in the same manner as found in the combinatorial studies. One objective of this chapter is to demonstrate that such a point does exist for bulk samples synthesized in either air or oxygen.

It is important to confirm that the combinatorial results hold up well when synthesis of bulk samples is done in a tank reactor. In this chapter, a few preliminary results obtained by Aaron Rowe will be included in order to discuss how boundaries change for bulk samples, particularly those heated to higher temperatures.

Data in this chapter are reprinted from Refs. [12] and [84] with permission from the American Chemical Society and the Electrochemical Society.

E. McCalla, *Consequences of Combinatorial Studies of Positive Electrodes for Li-ion Batteries,* Springer Theses, DOI 10.1007/978-3-319-05849-8_7,

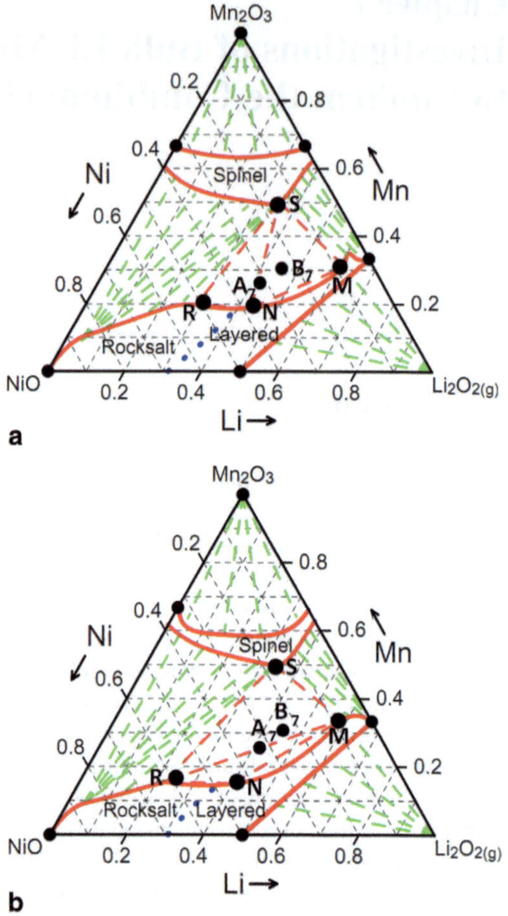

Fig. 7.1 The pseudo-ternary phase diagrams obtained for the Li–Ni–Mn–O system for combinatorial samples obtained by quenching (**a**), and slow cooling (**b**). *Red lines* represent boundaries to single-phase regions, *green dashed lines* are tie-lines, *red dashed lines* are tie-lines bounding three-phase regions and the *blue dotted line* is the phase transition between cubic and layered rocksalt structures. There is a *red tie-line* joining R and N, however the two-phase region is so small that it cannot be distinguished from the layered boundary on the scale shown here. Points *S*, *M*, *N* and *R* refer to the corners of the three-phase regions and move with synthesis conditions. Points A_7 and B_7 are two compositions referred to in the text

Finally, to determine the impact of the structures on the electrochemistry of the positive electrodes, coin cells of the M, N and R phases were made. These samples were the same quenched bulk samples heated in oxygen used in Chap. 6 to perform Rietveld refinements.

7.2 Experimental Design

Bulk samples M, N, R, A_7, A_7' and B_7 were made using the one-pot synthesis approach described in Sect. 2.1.2. Sample A_7' is not shown in Fig. 7.1, as it is very near to A_7, only a little closer to the nickel corner. The M, N and R materials were heated in oxygen to 800 °C for 5 h before quenching. Coin cells were made with these materials as outlined in Sect. 2.5. Cells were cycled at a rate of 10 mA/g. The

Table 7.1 Metal molar fractions for samples discussed in the text. For samples A_7, A_7' and B_7, all heated in oxygen, the composition is as-dispensed, while for C_7–G_7, heated in air, the compositions shown were obtained with elemental analysis for the regular cooled samples. The values for the M, N, R and S samples are for synthesis in oxygen from Chaps. 5 and 6

Sample	Li	Mn	Ni
A_7	0.425	0.255	0.32
A_7'	0.41	0.24	0.35
B_7	0.45	0.3	0.25
C_7	0.330	0.335	0.335
D_7	0.367	0.317	0.317
E_7	0.441	0.280	0.279
F_7	0.450	0.276	0.274
G_7	0.495	0.253	0.252
Quench			
M	0.61	0.31	0.08
N	0.40	0.17	0.43
R	0.29	0.20	0.51
S	0.36	0.50	0.14
Slow cool			
M	0.58	0.34	0.08
N	0.40	0.15	0.45
R	0.22	0.16	0.62
S	0.33	0.50	0.17

first two cycles were over the range 2.5–4.6 V, followed by ten cycles from 2 to 5 V. Finally the M material was tested for long-term cycling after the 12 preliminary cycles mentioned here. The long-term cycling was performed over the range 2–4.8 V.

The A_7, A_7' and B_7 samples were heated under various conditions discussed in the results section and characterized with X-ray diffraction (XRD) using the JD-2000 diffractometer.

Bulk samples C_7–G_7 were made by the tank reactor method (Sect. 2.1.2) and characterized with XRD by Aaron Rowe. These samples were heated to either 800 or 900 °C for 12 h in air and then either quenched or regular cooled. The XRD scattering from these samples was then measured in either the JD-2000 or the D-5000 diffractometer.

7.3 Structural Results

Figure 7.2 shows XRD scans from bulk samples made at the composition A_7 given in Table 7.1 and shown in Fig. 7.1. These scans demonstrate that the four-phase transformation does occur. Here, the progression of the phase transformation can be seen in the area of the R peak near 43.6° which grows steadily from scan (a) toward scan (e). The quenched samples, (a) and (b), show how rapidly the R phase grows with a significant amount present when quenched on a steel plate and none when liquid nitrogen was used. This rapid growth is consistent with the scanning electron microscopy (SEM) images from the previous chapter showing a small number of large cubic crystallites in a sample containing the N, M and R phases. Scan (f) in Fig. 7.2 shows that N replaced R when the slow cooled sample was reheated and quenched; consistent with the transformation being reversed when heated back to high temperature. However, there were still remnants of the R phase here suggesting that some lithium had been lost during the extended heating, moving point A_7

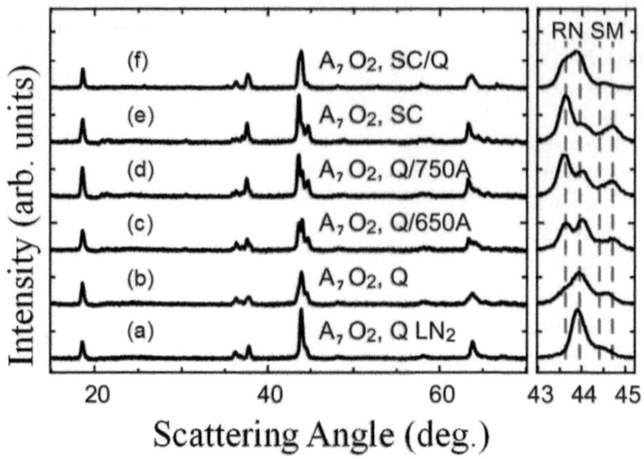

Fig. 7.2 X-ray diffraction (XRD) patterns obtained for bulk sample A_7 in the Li–Mn–Ni oxide system heated to 800 °C. Composition A_7 is indicated in Fig. 7.1. The right frame focuses on the XRD peaks near 44 ° where the four phases have one peak each. The *red dashed* lines represent the positions of the *N*, *R*, *S* and *M* peaks when slow cooled after heating in oxygen. *SC* denotes samples slow cooled at 1 °C/min, *Q* stands for quenching on steel while *Q LN2* refers to quenching in liquid nitrogen, and *SC/Q* indicates a slow cooled sample that was reheated to 800 °C and quenched. *Q/650 A* and *Q/750 A* were quenched samples that were annealed at 650 °C and 750 °C respectively for 5 h before quenching back to room temperature

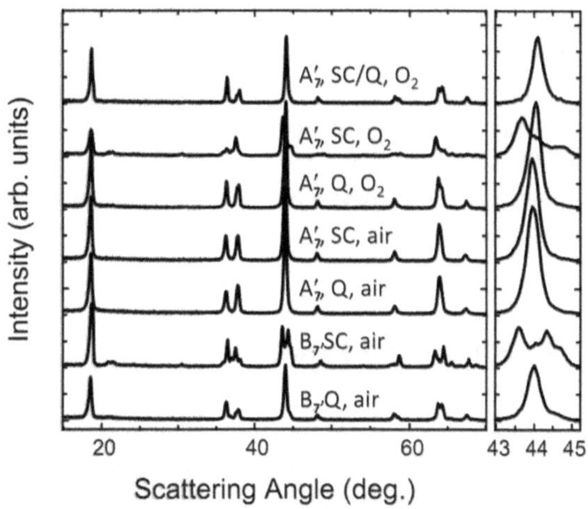

Fig. 7.3 X-ray diffraction (XRD) scans of samples made at compositions A_7' and B_7, indicated in Table 7.1. An expanded view of the peaks near 44 ° is included in the *right panel*. *Q* indicates quenched while *SC* refers to slow cooling (1 °C/min) and *SC/Q* denotes a slow cooled sample after being reheated to 800 °C and quenched

away from the lithium corner and into the RNS three-phase region or, once again, quenching on steel was not fast enough to prevent phase transformations entirely.

Figure 7.3 shows XRD scans of bulk samples at two compositions, A_7' and B_7, given in Table 7.1 and B_7 is also shown in Fig. 7.1. In oxygen, the quenched bulk sample at A_7' (Fig. 7.3) appears to be made up of the N phase only (though there may be traces of the S phase as well). This scan of a bulk sample quenched in oxygen suggests that the layered boundary has moved upward compared to that seen in the

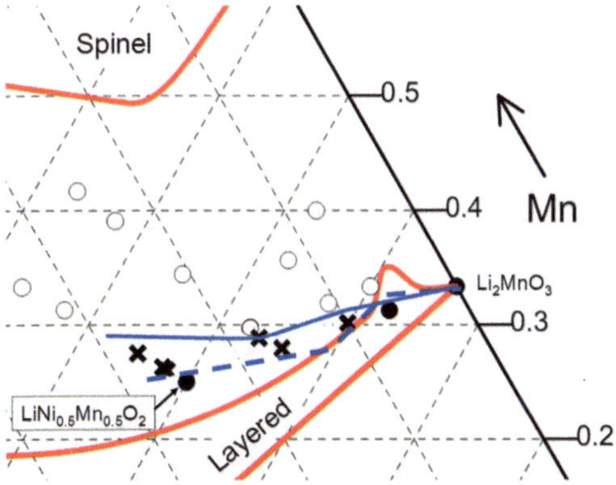

Fig. 7.4 The single-phase boundaries obtained in this study for quenched combinatorial samples heated in oxygen (*solid red lines*). The points were obtained by visual inspection of XRD scans of bulk samples synthesized in air; the *open circles* represent multi-phase samples for both quenched and regular cooled conditions, the *closed circles* are single-phase for both, and the x symbols correspond to single-phase quenched samples that phase separates during regular cooling. The *blue solid line* is a rough estimate of the boundary in air for quenched samples based on these data points, while the *blue dashed line* is for regular cooling. Further work to completely determine the phase boundaries in air is currently being done by Aaron Rowe

combinatorial samples such that A_7' lies near to the layered boundary. Upon slow cooling in oxygen, however, at least three phases are present in the bulk sample at A_7' consistent with the expected R, N and M phases. This behavior is consistent with the boundary moving down during slow cooling such that the sample is now in the RMN three-phase region shown in Fig. 7.1. The sample that was then reheated and quenched shows that this phase change was completely reversible demonstrating that the layered boundary moved back upon heating. Bulk samples of A_7 and A_7' together therefore show that the four-phase equilibrium exists and the transformations during cooling are reversible.

By contrast, both scans of sample A_7' obtained in air appeared to consist of a single phase in Fig. 7.3. This is consistent with the boundary having moved even higher in air. To confirm that the coexisting regions do exist in air, quenched and slow-cooled bulk samples at composition B_7 were prepared. Figure 7.3 shows that the quenched sample shows at least two phases (N and M with perhaps S causing an asymmetry in the N peak) consistent with the NMS triangle expected for quenched samples. Upon slow cooling, the XRD scan showed that conversion toward R, S and M was taking place but some N phase remained, implying that equilibrium was not reached and four peaks were seen in the XRD pattern. These few bulk samples show that although the boundaries move quite dramatically depending on synthesis conditions, the phase diagram shows the same features as those seen in the combinatorial studies. Hence, the current study should help considerably in interpreting data collected under various synthesis conditions.

Figure 7.4 shows preliminary results obtained for the tank reactor bulk samples made in air at 800 °C along with the boundaries obtained in oxygen in the combinatorial studies. These results are included here to show how the layered boundary moves

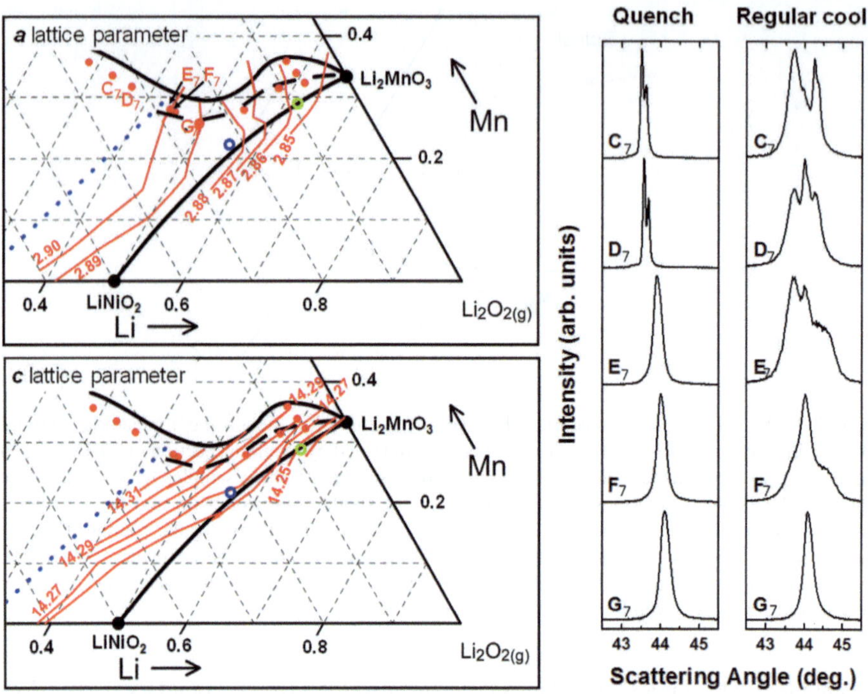

Fig. 7.5 A partial Li–Mn–Ni oxide phase diagram for bulk samples synthesized in air at 900 °C. The *thick solid black lines* are estimates to the boundaries of the layered region when quenched, the *dashed black line* is the upper boundary when slow cooled, and the *blue dotted line* is the cubic to layered phase transition. The contour plots shown as *thin red lines* for the lattice parameters were obtained using both combinatorial samples (not shown, from Chap. 6) and bulk samples (*red points*); all plotted compositions were determined by elemental analysis (with an uncertainty of about 0.02 in molar fraction)

upward as oxygen partial pressure decreases, particularly for quenched samples. The behavior near the Li–Mn line (where a sharp bump was obtained in oxygen) will be discussed in detail in the next chapter. Again the boundary was found to move downward during slow cooling consistent with oxygen content increasing in the samples thereby favouring the spinel phases.

Figure 7.5 shows the results for bulk samples in the Li–Mn–Ni–O system heated to 900 °C in air. These conditions are similar to those used in commercial synthesis of positive electrodes. The phase diagram is similar to that obtained with the combinatorial samples in oxygen though the solid-solution boundaries shift with the decrease in oxygen partial pressure and increase in temperature. The XRD patterns in the range 43–46 ° show the phase transformations taking place during cooling. Samples C_7–G_7 were all single-phase when quenched while samples C_7–F_7 showed multiple phases when cooled more slowly. Sample E_7 showed four phases when regular cooled, again suggesting that conversion to the three new phases did not reach completion so that equilibrium was not reached at intermediate cooling rates.

The two peaks seen in the quenched C_7 and D_7 samples correspond to the $K\alpha_{1,2}$ splitting visible due to large crystallites giving sharp peaks and do not indicate phase separation. The XRD results were used to create the approximate boundaries shown in Fig. 7.5. The upward movement of the layered boundary with temperature can be attributed to the reaction equilibrium favoring the layered phase over the spinel phase that contains more oxygen per metal atom. This again occurs as the high entropy of the oxygen gas wins out at elevated temperatures. The improved performance of the Li-rich materials synthesized at high temperature [2, 86] may therefore simply be due to the samples remaining single-phase, thereby avoiding the layered–layered coexistence region. This will be examined in detail in Chap. 9.

Furthermore, the layered region extends quite low in the Gibbs triangle compared to solid-solutions known prior to this thesis (Fig. 1.8b in the introduction). This implies that some published results must be reexamined. For example, a sample reported by Ohzuku et al. [52] (blue point in Fig. 7.5) had a very high capacity and is reported as being $LiNi_{0.5}Mn_{0.5}O_2$ which is very near to point G_7. However, the contour plots were used along with the published lattice parameter values of $a = 2.883$ Å and $c = 14.269$ Å to position this sample on the phase diagram in Fig. 7.5. This result suggests that the sample lost very little of the 25 % excess lithium used during synthesis such that the material was in fact a lithium-rich layered material, and thus its improved electrochemistry over other published $LiNi_{0.5}Mn_{0.5}O_2$ samples can be attributed to the extra lithium after nickel oxidation. Similarly, the open green symbol in Fig. 7.5 represents a sample from Ref. [51] that was reported to lie on the Li-rich line joining Li_2MnO_3 to $LiNi_{0.5}Mn_{0.5}O_2$. This point was positioned using the published lattice parameter, $a = 2.860$ Å and $c = 14.246$ Å, again suggesting that it too was below the Li-rich line such that not all of the 5 % excess lithium was lost during synthesis. These two points were used to sketch the curved line joining $LiNiO_2$ to Li_2MnO_3. The electrochemistry of single-phase layered materials below the lithium-rich layered line must be carefully studied and elemental analysis is required to confirm final compositions of samples in this region of the Gibbs triangle.

7.4 Electrochemistry of the R, M and N Phases

Figure 7.6 shows the electrochemical data obtained for the three new materials: N, M and R made using the one-pot synthesis approach and heated to 800 °C in oxygen for 5 h before being quenched. The only material with promising electrochemical properties was the M-layered phase. This is not surprising, since the XRD patterns in Chap. 6 showed this material to be a highly ordered lithium-rich material. While some of the lithium was removed by oxidizing the nickel from 2+ to 4+ up to 4.45 V [2], the remaining lithium was removed at about 4.5 V via the high voltage plateau discussed in the introduction. This plateau has long been thought to involve lithium being removed along with oxygen gas via the so-called oxygen release process [2, 17]. This process would leave the oxidation states of the transition metals unchanged

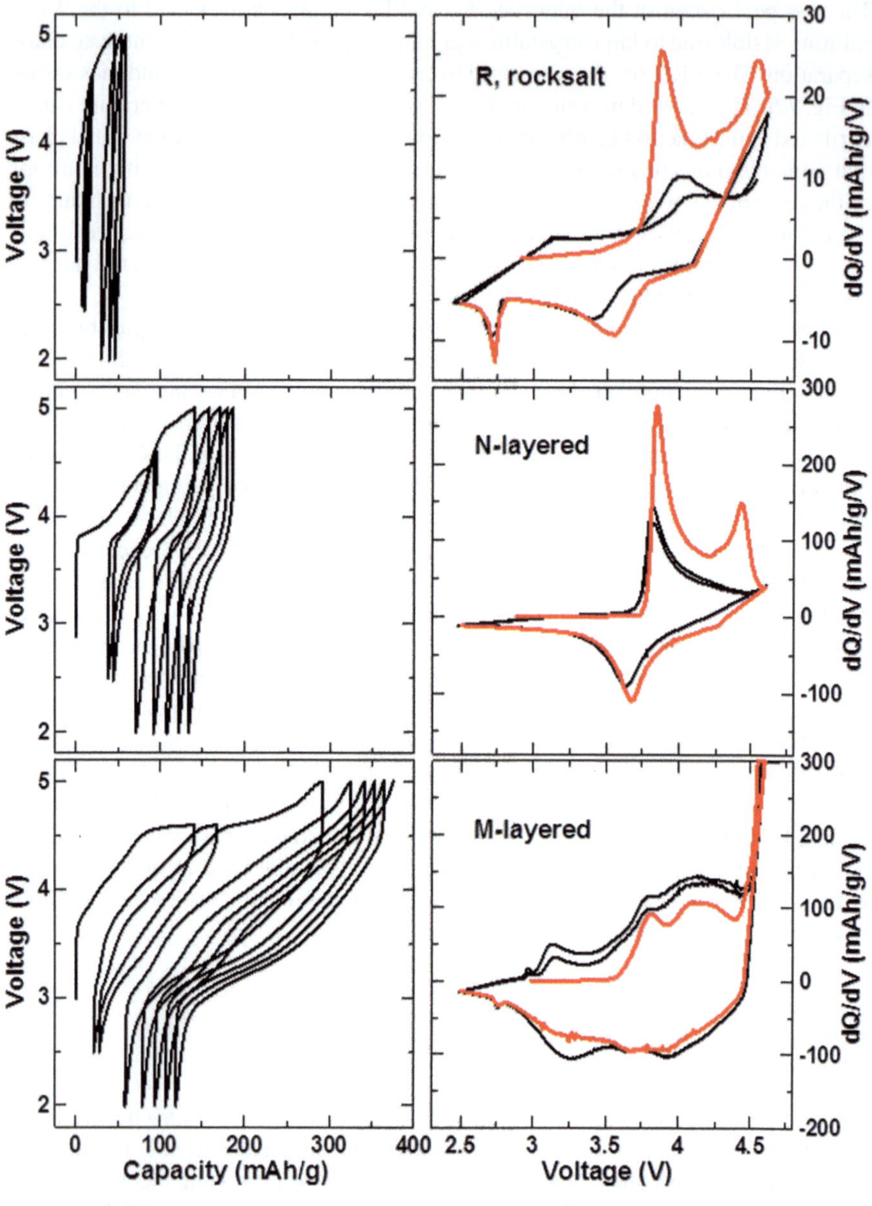

Fig. 7.6 Electrochemical data of half cells made from the M, N and R materials obtained by heating in oxygen at 800 °C and quenching to room temperature. The cells were first cycled between 2.5–4.6 V for two cycles (shown in all panels, with the first cycle shown as a *thick red line* in the dQ/dV plots), then between 2–5 V for ten cycles (for clarity, only a few of these are included in the voltage versus capacity plots). All measurements were performed at a constant specific current of 10 mA/g such that the M-layered material was cycled at a rate of approximately C/24 (i.e., 24 h to charge or discharge)

such that the manganese could then be reduced from its initial 4+ state during the next discharge, thereby activating it. More recently, Koga et al. [15] and Sathiya et al. [16] have demonstrated oxygen participation in the redox process in some lithium-rich oxides—a process that is accompanied with a phase transformation. Koga demonstrated that Li–Co–Mn–Ni–O layered materials with excess lithium convert to a two-phase material after the high voltage plateau, and this was attributed to the shell of particles losing oxygen while the core did not, such that much of the high voltage plateau can be attributed to oxygen redox in the core of the particles. Fell et al. [87] recently found this two-phase behavior in $Li_{1.2}Ni_{0.2}Mn_{0.6}O_2$ and used Rietveld refinement to determine that one of the phases was oxygen deficient while the other was not. At this moment, it is therefore unclear how much oxygen is lost and how much oxygen redox takes place in the M material discussed here during the high voltage plateau and remains an important question to answer. Regardless of the ongoing debate about the nature of the high-voltage plateau, both the nickel redox and the high voltage plateau can be seen as peaks in the dQ/dV curve during the first charge of the M material in Fig. 7.6 The new peak appearing at 3.1 V in the second cycle was consistent with manganese redox. After the cycles shown in Fig. 7.6, the M phase was tested between 2–4.8 V to test its long-term cycling performance. The long-term cycling capacities were 245.0 mAh/g during charge and 236.5 mAh/g during discharge initially, and 244.2 mAh/g and 237.4 mAh/g respectively 40 cycles later, demonstrating stable cycling.

By contrast to the highly ordered M-layered phase, the N-layered structure has a disordered lithium layer with 30 % nickel occupation that apparently prevents some lithium from diffusing out of the material, since only 50 % of the 201 mAh/g total theoretical capacity was obtained during the first charge shown in Fig. 7.6. The capacity also faded rapidly such that this material is not attractive for Li-ion batteries. The cubic rocksalt phase, R, did have some ordering of lithium, manganese and metal site vacancies on the cubic lattice as discussed in Chaps. 5 and 6. Nonetheless, a capacity of only 20 mAh/g was achieved on first charge even though the theoretical capacity was 138 mAh/g, demonstrating that most lithium diffusion paths were blocked or severely hindered.

A consequence of the poor performance of the N and R phases is that synthesis conditions and compositions that give rise to these phases must be avoided. This limits the region of interest for layered–spinel composite electrodes to the area to the right of the M–S line in Fig. 7.1 (b), with the M–S line being of particular interest since both the M and S phases are decent electrode materials. Furthermore, since the R phase appears in both three-phase regions during slow cooling, the effect of quenching on composite electrodes must be studied. The knowledge gained about the phase diagram also affects core-shell materials [88], where understanding the equilibrium phases is necessary to determine whether or not ion mobility at a given temperature is sufficient to destroy the desired core-shell structure.

The poor performance of the N-phase also impacts the so-called layered–layered nano-composites that have long been searched for along the Li-rich layered line with inconclusive results [44, 46]. The phase diagram suggests that samples showing this short range phase separation should appear elsewhere in the triangle, namely in the region directly above point G_7 in Fig. 7.5. During slow cooling, the upper boundary of the layered region moves downward. This implies that the layered–layered region also moves downward as the M and N points move. A sample made near the composition of $LiNi_{0.5}Mn_{0.5}O_2$ (point G_7) would be single phase if quenched from high temperature and would transform toward a layered–layered composite when regular cooled. The endpoints of the phase separation would lie along the boundary between N and M. Since the N phase contains a large amount of nickel on the lithium layer, the formation of nano-composites in this composition space would involve the clustering of nickel on the lithium layer which could interfere with lithium diffusion. This well-known composition, $LiNi_{0.5}Mn_{0.5}O_2$, therefore warrants further study and will be discussed from this new point of view in Chap. 9.

7.5 Conclusions Based on Bulk Li–Mn–Ni–O Samples

The key features in the Li–Mn–Ni–O pseudo-ternary phase diagrams, as determined using combinatorial samples, were confirmed with bulk samples synthesized under various conditions via two different synthesis routes. The four-phase equilibrium was observed and transformations were found to be reversible, thereby confirming that two three-phase regions exist and transform during slow cooling. The primary importance of this work is a better understanding of how the phase boundaries and coexisting regions transform when cooled at rates typically used commercially. It is also important to recognize that a small amount of transformation occurred even for samples quenched on a copper or steel plate. These changes cannot be avoided entirely without quenching in liquid nitrogen or avoiding the compositions where the transformations occur. Understanding the phase diagrams should have a significant impact on research focused on composite electrodes in the Li–Mn–Ni–O system, since this work identified the compositions and conditions required to obtain layered–layered nano-composites and layered–spinel composites.

The lattice parameter contour plots developed here also showed that some samples made with excess lithium retain most of the excess, even after heating to high temperatures. Elemental analysis on a sample in the layered region is therefore mandatory in order to be confident of the composition of the final product.

Chapter 8
Layered Materials with Metal Site Vacancies

8.1 Motivation for the Study of Samples near Li_2MnO_3

In the Li–Mn–Ni–O system, studies of the lithium-rich layered materials have generally been limited to the composition line between Li_2MnO_3 and $LiNi_{0.5}Mn_{0.5}O_2$ (referred to here as "stoichiometric lithium-rich") where reversible cycling above 250 mAh/g has been achieved [2, 3, 43]. These materials take the O3-type structure with lithium layers containing a small amount of nickel and transition metal (TM) layers being made up of manganese, nickel, and some lithium. The oxidation states of nickel and manganese are typically 2+ and 4+ in these structures [90]. In the previous chapters, it was found that the solid-solution layered structures can be made with either more or less lithium than the stoichiometric lithium rich line.

Figure 8.1 shows the phase diagram obtained by quenching combinatorial samples in oxygen. Also shown are composition lines obtained by keeping nickel and manganese oxidation numbers constant. This diagram shows that all spinel samples are consistent with Ni^{2+} and $Mn^{3,4+}$. The rocksalt/layered lines show that the majority of samples in this region have Mn^{4+} with $Ni^{2,3+}$. That excess lithium can be added to the stoichiometric lithium-rich materials while keeping the structures single-phase has been well known for some time and has been used extensively to make high capacity materials [51, 52]. These structures can be understood as being identical to the usual Li-rich materials with some of the nickel oxidized to 3+ during synthesis, which is quite feasible given that $LiNiO_2$ can be made under the same conditions. However, there are two exceptions in Fig. 8.2 where layered materials cannot be made without reducing some manganese to 3+ or having metal site vacancies: the "bump" region near the Li–Mn line (to be discussed here) and the top of the ordered rocksalt region. Structures in the ordered rocksalt region have already been shown to contain metal site vacancies which allow a higher manganese oxidation state (perhaps even keeping it in the 4+ state), as discussed in Chaps. 5 and 6. However, for the bump region, the ambiguity remains: there may be metal site vacancies and/or some manganese reduced to 3+. This ambiguity requires further study and will be resolved in this chapter.

Data in this chapter are reprinted from Ref. [89] with permission from the American Chemical Society.

E. McCalla, *Consequences of Combinatorial Studies of Positive Electrodes for Li-ion Batteries,* Springer Theses, DOI 10.1007/978-3-319-05849-8_8,

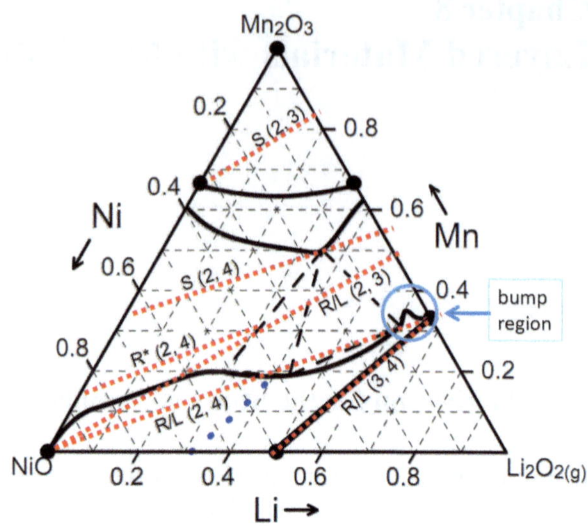

Fig. 8.1 The phase diagram obtained in oxygen by quenching. The *red dotted lines* represent constant oxidation number lines, *S* refers to spinel compositions $Li_xMn_yNi_{3-x-y}O_4$, *R/L* denotes rocksalt or layered structures $Li_xMn_yNi_{2-x-y}O_2$, and *R** denotes ordered rocksalt structures with unoccupied 4b sites. The coordinates used to label the lines are: (Ni oxidation number, Mn oxidation number)

Figure 8.2 shows an approximate phase diagram of samples made in air at 900 °C and then quenched, included in the previous chapter. Again, in the combinatorial studies, the single-phase material in the bump region with the most lithium deficiency had the metallic molar composition: $Li_{0.6}Mn_{0.35}Ni_{0.05}$. The two models considered here (metal site vacancies or Mn^{3+}) would predict the following two compositions: $Li[Li_{0.136}Ni^{2+}{}_{0.146}Mn^{3+}{}_{0.154}Mn^{4+}{}_{0.564}]O_2$ or $Li[Li_{0.057}Ni^{2+}{}_{0.136}\square_{0.139}Mn^{4+}{}_{0.668}]O_2$ where □ indicates vacant metal sites. The main objective of this chapter is to determine which of these two models is correct. Given that the vacancy model predicts that 6.95 % of the metal sites remain vacant, density measurements as well as redox titrations are accurate enough to distinguish between the two models.

In the combinatorial study, the compositions at which one can expect layered–layered materials to appear were identified. It is of interest that the bump region discussed here never underwent such a phase separation. Instead, the upper section of the bump phase separated into layered–spinel composites with the region near the stoichiometric lithium rich line remaining single phase. To better understand why this is the case, this chapter includes a Monte Carlo simulation of a material in this bump region. The details of this simulation are given in Sect. 2.12.

8.2 Experimental Design

Samples A_8–D_8, indicated in Fig. 8.2, were synthesized by Aaron Rowe using the tank reactor method and were heated to 900 °C for 12 h before quenching between copper plates.

Li_2MnO_3 was made by John Camardese as a reference material for this study by solid state synthesis. Lithium carbonate, with 10 % excess, was mixed with manganese oxyhydroxide (MnOOH, Chemetals) and heated to 1000 °C for 48 h.

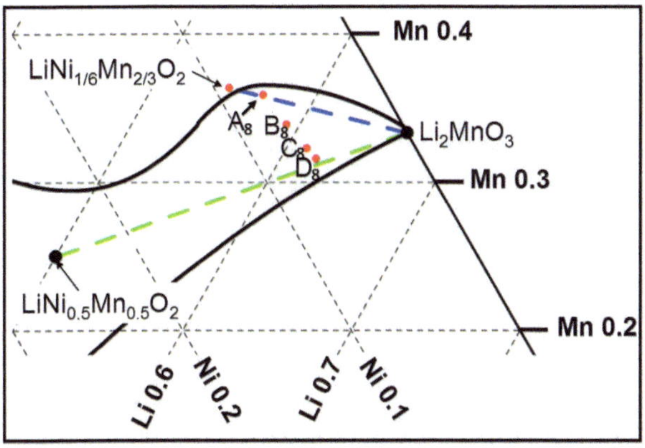

Fig. 8.2 A partial phase diagram for bulk samples quenched from 900 °C, with compositions indicated for samples discussed throughout this chapter. The *black boundaries* are estimates to the boundaries of the layered region, while the *blue dashed line* is the composition line with sufficient vacancies to have Ni^{2+}, Mn^{4+} and 2/3 of TM filled with Mn^{4+} corresponding to the solid-solution series $Li[Li_{1/3-x}Ni_{x/2}\square_{x/2}Mn_{2/3}]O_2$ with $0 \leq x \leq 1/3$. The *green dashed line* is the stoichiometric lithium rich line discussed in the text

A sample of $Li[Ni_{1/6}\square_{1/6}Mn_{2/3}]O_2$ (approximately 2 g) was made using the one-pot synthesis method described in Sect. 2.1.2. The sample was heated for 5 h in air at 900 °C and then quenched by transferring it onto a copper plate.

For samples A_8–D_8 and $Li[Ni_{1/6}\square_{1/6}Mn_{2/3}]O_2$, inductively coupled plasma was used as elemental analysis to determine the actual metal molar fractions after heating. The X-ray scattering from all samples was measured in a Siemens D-5000 diffractometer. The resulting scans were then analyzed using Rietveld refinement. The refined value for the occupation of the oxygen sites converged to above 100 % such that it was used to calculate the vacancy content on the metal sites.

The true density of the samples was measured using a helium pycnometer. With values for lattice parameters and elemental analysis, the density of the samples were used to calculate the occupancy of the metal sites directly. This gave sufficiently precise results since the measured sample density was as much as 6.3 % lower than the crystallographic density.

Redox titrations, as described in Sect. 2.9, were performed in order to determine the average oxidation states of manganese and nickel. Finally, with elemental analysis results, the average oxidation states were used to calculate the vacancy concentrations assuming all oxygen was in the (2−) state.

To confirm the results of the redox titrations, the manganese K-edge XANES spectra of samples A_8–D_8, Li_2MnO_3 and Mn_2O_3 were collected by Paul Duchesne.

Electrochemical tests were carried out on the $Li[Ni_{1/6}\square_{1/6}Mn_{2/3}]O_2$ sample using a standard 2325 coin cell design as described in Sect. 2.5.

Finally, the Monte Carlo simulation discussed in Sect. 2.12 was used to illustrate the structures of interest and to better understand the experimental results. All compositions used in the simulations were consistent with the elemental analysis results,

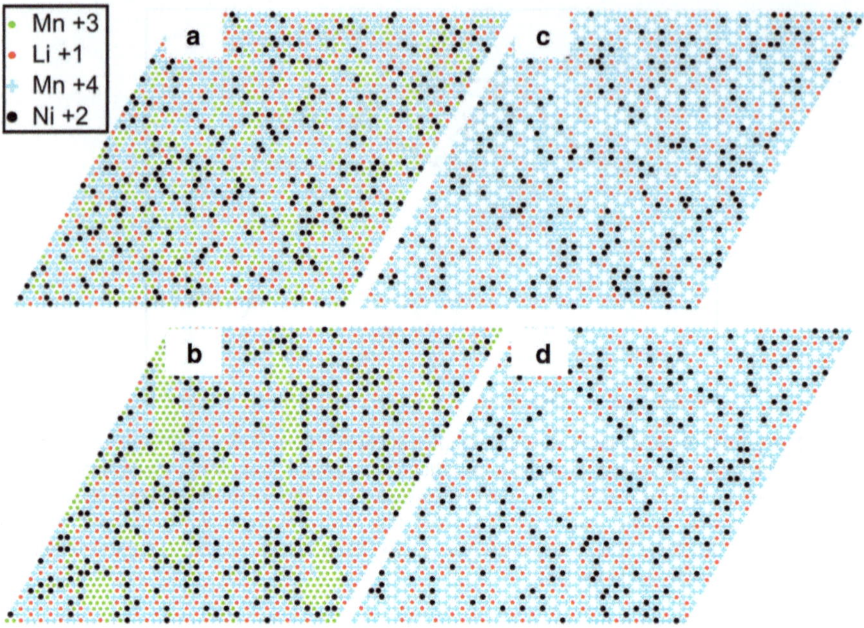

Fig. 8.3 Monte Carlo results of sample $Li_{0.6}Mn_{0.35}Ni_{0.05}$: the composition at the top of the "bump" from the combinatorial studies, as shown in Fig. 8.2. Two possible structures are included: (**a, b**) Mn^{3+} is present and (**c, d**) metal site vacancies (*white areas*) exist in sufficient concentrations to maintain Mn^{4+}. The diagrams shown were obtained at high temperature with $\beta_T = 0.5$ for **a, c** and a temperature ten times lower with $\beta_T = 5.0$ in **b, d**

though the calculations assumed no nickel on the lithium layer such that only the transition metal layers needed to be considered. The results from Chap. 4 suggested that $\beta_T = 1$ corresponds to a temperature where phase separation takes place. This temperature was experimentally determined to be below 800 °C for Li–Co–Mn–O materials. In the previous chapter, such transformations were found to occur at or below 750 °C in the Li–Ni–Mn–O system at certain compositions. Here, this same Monte Carlo simulation was used to explore the two possible structures at the top of the bump (point A_8 in Fig. 8.2) as well as a sample lower in the layered region (point D_8) in order to better understand why materials in the bump region do not form layered-layered composites.

8.3 Monte Carlo Results

Figure 8.3 shows the results of the Monte Carlo simulation for the two possible structures at the composition of the top of the bump: (Li, Mn, Ni) = (0.6, 0.35, 0.05). The two models are: (a) some manganese is in the0 3+ state and (b) there are metal site vacancies. For simplicity, the vacancies were all assumed to be on

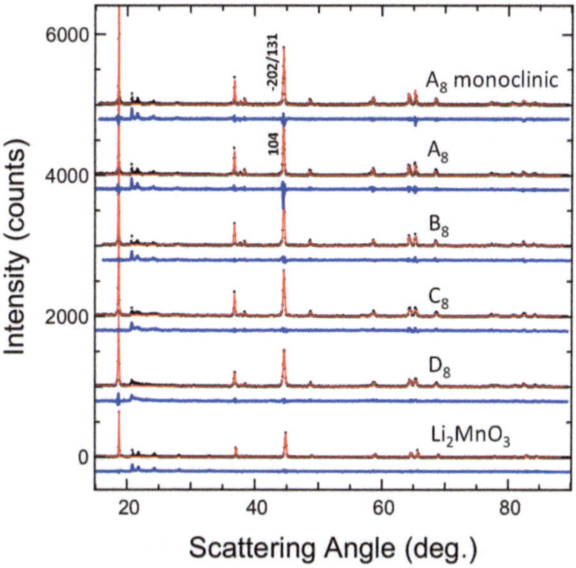

Fig. 8.4 XRD scans with fits obtained with Rietveld refinement (*red*) and difference plots (*blue*). All are fit as layered, except for the top scan which is fit as C/2m (monoclinic). Scans are offset vertically for clarity

the transition metal layer. However, if some vacancies were on the lithium layer the results of the Monte Carlo simulation would not change significantly since the roles of lithium and vacancies on the transition metal layers appear to be identical. Clearly, the presence of vacancies helps stabilize the structure, with manganese forming two-thirds of the transition metal layer. Without vacancies, manganese (III) had a tendency to cluster suggesting that such a structure would not form a solid solution particularly after slow cooling. The presence of vacancies in the model promoted a solid solution with random occupation of lithium, nickel, and vacancies on one $\sqrt{3} \times \sqrt{3}$ superlattice while only manganese occupied the other two superlattices. This structure suggests that the ordering peaks in the X-ray diffraction (XRD) should be sharp for this material. The fact that this composition never phase separated into two layered structures during the Monte Carlo simulation is consistent with experimental results where a sample near the top of the bump phase separated into layered–spinel composites. Since the Monte Carlo simulation does not allow for tetrahedral sites, a transformation involving spinel structures is not possible with the simple model used here.

8.4 Vacancy Measurements

Figure 8.4 shows the XRD patterns obtained for samples A_8–D_8 and Li_2MnO_3 along with fits obtained using Rietveld refinement assuming a layered R-3m structure. For sample A_8, the difference plot shows significant errors particularly near the (104) peak, so the scan was refined again assuming a monoclinic structure with the

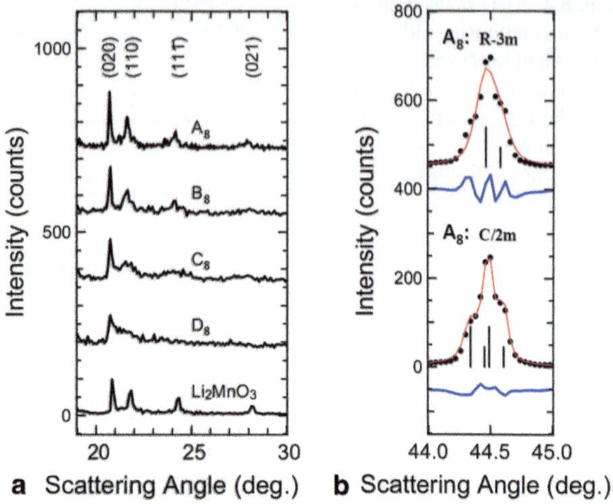

Fig. 8.5 **a** XRD patterns in the region where superlattice peaks appear. **b** Peak near 44.5° degrees in sample A_8 fit as both layered (*top*) and monoclinic (*bottom*) with scattering angle steps of 0.02°. Vertical lines in R-3m (*top*) correspond to $K\alpha_1$ and $K\alpha_2$ for the (104) peak, and to the (−202) and (131) peaks for C/2m (*bottom*)

C/2m space group. Figure 8.5b shows this region in more detail and clearly shows that monoclinic distortions have resulted in an asymmetry in the R-3m (104) peak such that a high quality fit can only be obtained using the monoclinic phase. The monoclinic fit yielded 9.3 ± 2.5 % vacancies; the larger uncertainty arising as it becomes difficult to constrain the extra parameters available in the monoclinic fit. Still, this value is consistent with the other vacancy measurements for sample A_8 discussed below. Furthermore, the lattice parameters were $a = 4.953(1)$ Å, $b = 8.574(2)$ Å, $c = 5.050(1)$ Å, and $\beta = 109.33(2)°$. The value for β is slightly larger than the value of 109.1° for a hexagonal lattice, thereby confirming that monoclinic distortions existed in this material, they were relatively small and were comparable to that seen in Li_2MnO_3 [24].

In order to confirm that the highly ordered structure seen in the Monte Carlo simulation is the stable phase found experimentally, the fraction of metal sites which are vacant have been determined for samples A_8–D_8 three different ways using Rietveld refinement, density measurements, and average metal oxidation state. Table 8.1 shows the results for the elemental analysis while Table 8.2 shows the results for the vacancy fraction measurements. The uncertainty in the density is a statistical error in the mean of multiple values. The only measure of absolute error presented here is the small error of 0.014 g/mL found for Li_2MnO_3, which is expected to have no vacancies. The calculated vacancy concentrations for Li_2MnO_3 show the precision of the methods used with both density measurements and the redox titration giving values very close to 0 % vacancies. For samples A_8–D_8, the agreement between the three approaches is generally excellent. The average value of 6.9 % with a standard deviation of 0.07 % for sample A_8 agrees with the expected 6.9 % required to keep all manganese in the 4+ state as discussed in Sect. 8.1. This helps confirm the structure generated with the Monte Carlo simulation where the two-thirds manganese occupation of sites on the TM layer allows for ordering of manganese and the formation of a solid solution on the remaining one-thirds of sites.

Table 8.1 Metal molar fractions for samples A_8–D_8 and Li_2MnO_3 discussed in this chapter. For Li_2MnO_3 the expected composition is shown, while for A_8–D_8 the actual compositions obtained by elemental analysis are shown

Sample	Li	Mn	Ni
A_8	0.568	0.359	0.073
B_8	0.592	0.339	0.069
C_8	0.612	0.323	0.065
D_8	0.621	0.316	0.063
Li_2MnO_3	0.666	0.333	0

Table 8.2 Results for Rietveld refinement, pycnometry, and redox titrations

Property	A_8	B_8	C_8	D_8	$Li_2MnO_3^b$
XRD					
a (Å)	2.8582 (1)	2.8558 (2)	2.8542 (2)	2.8536 (2)	$a=$ 4.9318 (3)
c (Å)	14.298 (1)	14.283 (2)	14.255 (3)	14.253 (3)	$b=$8.5375 (5)
					$c=$5.0299 (3)
					$\beta=$109.32 (1)°
Ni_{Li} (%)	3.1 (3)	2.1 (2)	2.3 (2)	2.7 (2)	—
vacancies (%)	6.9 (1.8)	8.1 (1.3)	2.7 (1.3)	1.0 (1.5)	—
Pycnometry					
density (g/mL)	4.138 (14)	4.089 (9)	4.0884 (12)	4.051 (7)	3.868 (7)
vacancies (%)	7.0 (5)	5.2 (4)	2.2 (5)	2.0 (5)	0.6 (3)
Redox titration					
Mn	3.99 (1)	3.98 (1)	4.00	4.00	3.99 (1)
Ni[a]	2.00	2.00	2.05(5)	2.20(5)	—
vacancies (%)	6.82 (16)	3.81 (16)	1.83 (15)	1.17 (15)	0.17 (17)

[a] The average nickel oxidation state was assumed to be 2.0 unless the manganese value was measured to be greater than 4.0. In the later case, the manganese oxidation state was assumed to be 4.0 with the nickel state > 2.0

[b] The Li_2MnO_3 XRD scan was fit as monoclinic, ignoring peaks between 20 and 33° and the crystallographic density assuming no vacancies is 3.882 g/mL

The results of the redox titrations show that manganese consistently had an oxidation state of 4+ in these materials. This is in contrast to recent results published by Simonin et al. [48] who report an average manganese oxidation state of 3.75 in a lithium-rich layered oxide. Figure 8.6 shows the Mn K-edge XANES spectra for samples A_8–D_8 as well as Mn_2O_3 and Li_2MnO_3 which were used as references for Mn^{3+} and Mn^{4+} respectively. The leading edge for each of the samples A_8–D_8 were extremely close to that of Li_2MnO_3 in relation to Mn_2O_3. This confirms the results of the redox titration.

Figure 8.5a shows that the superlattice ordering peaks get sharper from sample D_8 to A_8 (i.e., compositions moving upwards in the bump region). The superlattice peaks of sample A_8 were, in fact, as sharp as those of Li_2MnO_3 even though the latter was heated for far longer at a higher temperature. The superlattice peaks therefore support the claim that the vacancies allow for ordering of manganese on two-thirds of the TM layer. For samples B_8–D_8, the vacancy fraction diminishes as lithium content

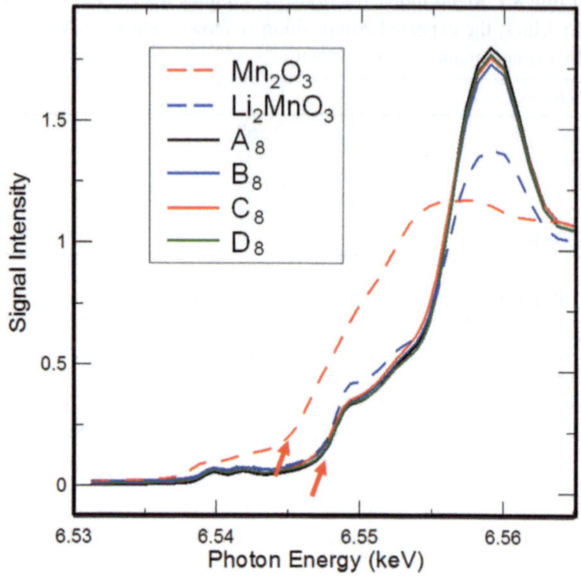

Fig. 8.6 Mn K-edge XANES patterns collected for samples showing metal site vacancies, along with Mn_2O_3 as a reference for Mn^{3+}, and Li_2MnO_3 for Mn^{4+}. The *red arrows* indicate the positions of the absorption edge near 6.545 keV for Mn_2O_3 and 6.548 keV for all other samples

increased but it never reached zero. Even sample D_8, which was on the stoichiometric lithium-rich line, had at least 1 % of the metal sites vacant.

It is also important to recognize that although the data provided here clearly demonstrates the existence of the metal site vacancies, the data here cannot be used to determine the locations of the vacancies. For example, the Rietveld refinement performed on the XRD scattering from sample A_8 resulted in quality factor values of $R_B = 3.74\,\%$, $R_P = 19.00\,\%$, $R_{WP} = 5.31\,\%$ assuming vacancies were spread evenly between the lithium and TM layers, while $R_B = 3.78\,\%$, $R_P = 19.02\,\%$, $R_{WP} = 5.33\,\%$ was obtained if vacancies were assumed to be on TM layers only. This small change in goodness of fit factors implies that X-ray diffraction cannot easily distinguish between a vacancy and a lithium atom. In fact, all experimental data here can only establish the average metal site occupations and not the locations of the vacancies, such that studies that are sensitive to lithium atoms, such as neutron scattering, are warranted in order to determine whether or not the vacancies are limited to the transition metal layer.

The local structure of sample D_8, which lies on the stoichiometric lithium-rich line, must now be reconsidered. Up to now, it has been assumed that structures on this composition line have Ni^{2+} and Mn^{4+} only. However, Table 8.2 shows that this is not the case: there is some Ni^{3+} as well as about 1 % of the metal sites vacant. Figure 8.7 shows the results of a Monte Carlo simulation with the following composition on the TM layer: ${Li^+}_{0.227}{Ni^{2+}}_{0.100}\square_{0.023}{Ni^{3+}}_{0.025}{Mn^{4+}}_{0.625}$. At high temperatures ($\beta_T = 0.5$), the simulation gave a structure very similar to that seen in Fig. 8.3 with manganese occupying two superlattices; however, here Ni^{3+} substituted for manganese on these sites. Together, Mn^{4+} and Ni^{3+} occupy 65 % of the TM layer, thereby allowing them to very nearly occupy two $\sqrt{3} \times \sqrt{3}$ lattices. It is

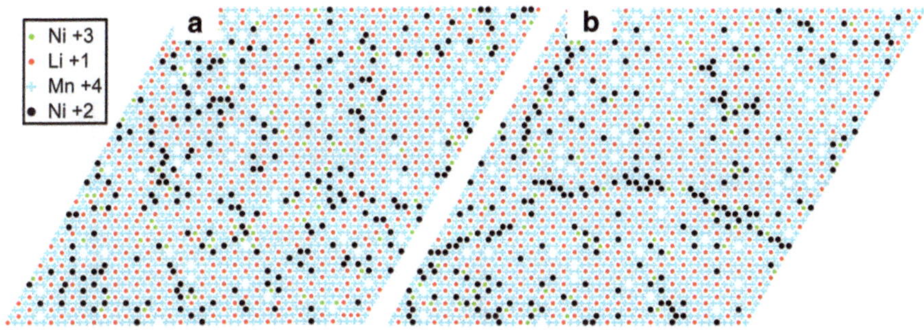

Fig. 8.7 Results of a Monte Carlo simulation for sample D_8 using the vacancy content and average nickel oxidation state from Table 8.1. The structures were obtained with $\beta_T = 0.5$ (**a**) and $\beta_T = 2.0$ (**b**)

also interesting to note that Ni^{2+} had a tendency to cluster in the simulation at lower temperature ($\beta_T = 2$, though this same structure was seen consistently for $\beta_T > 1$). This clustering appears to give rise to nickel-rich grain boundaries with the ordered vacancy-containing structure forming the bulk of the crystallite. In ab-initio calculations, Hinuma et al. [55] found that an arbitrary energy penalty on Ni–Ni nearest neighbor interactions was necessary to prevent clustering of nickel on the lithium layer in $LiNi_{0.5}Mn_{0.5}O_2$. The simple Monte Carlo simulation may simply be encountering the same issue and this clustering may not occur. Either way, experimental confirmation is necessary and the Monte Carlo simulation at the very least suggests that a stable phase exists at high temperature where Mn^{4+} and Ni^{3+} can order on two superlattices and a random occupation of Li, Ni^{2+} and vacancies exists on the third.

8.5 $Li[Ni_{1/6}\square_{1/6}Mn_{2/3}]O_2$

An important consequence of the stable phase at point A_8 is that it suggests the existence of a solid-solution line linking Li_2MnO_3 to A_8. This line would represent layered structures with manganese in the 4+ oxidation state occupying two-thirds of the TM layer sites: $Li[Li_{1/3-x}Ni_{x/2}\square_{x/2}Mn_{2/3}]O_2$ with $0 \leq x \leq 1/3$. This solid solution would terminate at the composition $Li[Ni_{1/6}\square_{1/6}Mn_{2/3}]O_2$, or in terms of metallic fractions: $Li_{0.545}Mn_{0.364}Ni_{0.091}$. This is in good agreement with $Li_{0.542}Mn_{0.369}Ni_{0.089}$ obtained by chemical analysis for the sample made at this composition by heating at 900 °C for 5 h before quenching. Figure 8.8 shows the XRD pattern obtained for this sample, refined as hexagonal R-3m. The XRD pattern shows a trace amount of contaminant phase with a peak near 44° as seen in the difference plot. Based on the phase diagram in Fig. 8.2, this contaminant is most likely a spinel phase. Despite this, the material is very nearly single phase and the results of the Rietveld refinement are: $a = 2.8580(1)$ Å, $c = 14.311(2)$ Å, 9.0 ± 1.6 % metal site

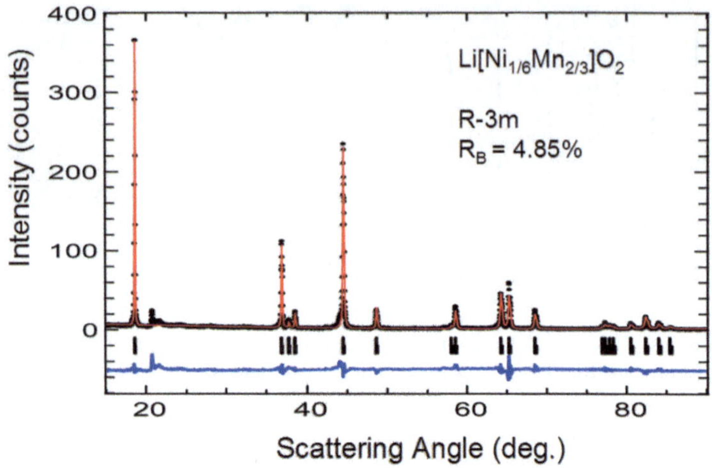

Fig. 8.8 XRD pattern of $Li[Ni_{1/6}\square_{1/6}Mn_{2/3}]O_2$ with the result of Rietveld refinement. *Vertical lines* are the calculated peak positions for $K\alpha_1$ (full length) and $K\alpha_2$ (half length)

vacancies, and 3.7 % nickel occupation on the lithium layer. The *a* lattice parameter is very close to that of sample A_8, while the *c* lattice parameter is noticeably larger, consistent with the contour plots provided in the previous chapter. Furthermore, the expected vacancy content for $Li[Ni_{1/6}\square_{1/6}Mn_{2/3}]O_2$ is 9.09 % in excellent agreement with the experimental value. It should be noted that with vacancies taken into account, this material is not lithium rich, lithium only occupies 50 % of the metal sites. The only lithium on the transition metal layer in this material comes about as a result of nickel disorder on the lithium layer; as is the case for other non Li-rich layered materials such as $Li[Ni_{0.5}Mn_{0.5}]O_2$.

Figure 8.9 shows electrochemical data obtained for the $Li[Ni_{1/6}\square_{1/6}Mn_{2/3}]O_2$ sample cycled at 10 mA/g at 30 °C. The usual features for Li-rich layered material can be seen in the dQ/dV plots such as the nickel redox up to about 4.45 V and the large irreversible peak referred to as the high voltage plateau just above 4.5 V. There is no sign of manganese 3+/4+ redox during the first cycle, again consistent with metal vacancies allowing for Mn^{4+} only in the starting material. However, with continued cycling the manganese redox peak near 3.1 V during discharge grew continuously. This transformation is typically attributed to conversion to spinel-like structures and results in a decrease in average voltage [91]. Since spinel structures have far more metal site vacancies in the transition metal layers than the layered structures, it is possible that the presence of vacancies on the TM layer in $Li[Ni_{1/6}\square_{1/6}Mn_{2/3}]O_2$ promotes conversion to spinel at high voltage. Furthermore, despite showing the high voltage plateau typically associated with lithium rich oxides, this material is not lithium-rich: only half of the metal sites are occupied with lithium.

Figure 8.10 shows the capacity as a function of cycle number for two cells of $Li[Ni_{1/6}\square_{1/6}Mn_{2/3}]O_2$. The capacity for the material cycled at 10 mA/g at room temperature plateaus at about 150 mAh/g which is a small fraction of the theoretical

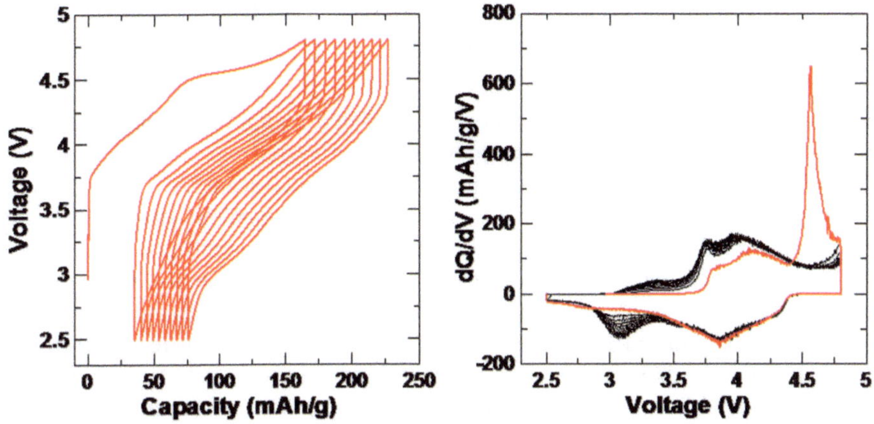

Fig. 8.9 Voltage versus capacity and dQ/dV plots for $Li[Ni_{1/6}\square_{1/6}Mn_{2/3}]O_2$ cycled at 30 °C with a specific current of 10 mA/g, or 0.020 mA/cm^2. The *red line* in the dQ/dV curve represents the first cycle

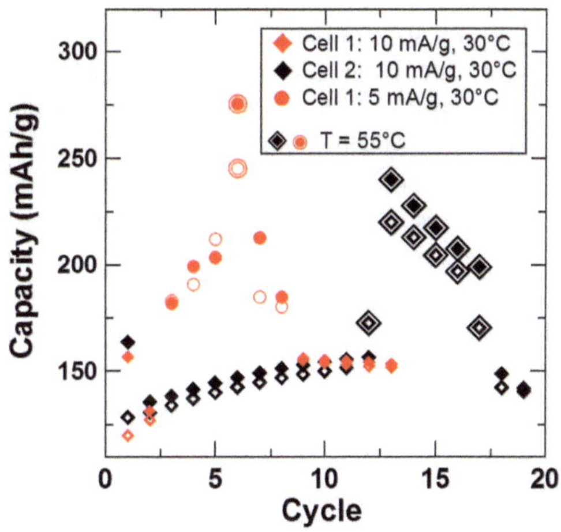

Fig. 8.10 Capacity versus cycle number for $Li[Ni_{1/6}\square_{1/6}Mn_{2/3}]O_2$. All *closed symbols* shown in the legend represent charging, while the *open symbols* are discharge capacities

capacity of 314 mAh/g assuming full lithium extraction. Higher capacities closer to the theoretical capacity can be achieved at slower rates and elevated temperatures. The maximum capacity obtained here was 250 mAh/g at 55 °C and a specific current of 5 mA/g. However, the capacity returned to about 150 mAh/g once the temperature and current were returned to 30 °C and 10 mA/g. These results show that a high temperature formation cycle cannot be used to increase the capacity, which is not competitive with state-of-the-art lithium-rich materials.

8.6 Conclusions Regarding Metal Site Vacancies in Li–Mn–Ni–O Materials

Layered Li–Mn–Ni–O materials were analyzed in a region of the phase diagram where a strange bump was seen in the boundary of the layered region. These structures were found to contain a significant amount of metal site vacancies. The maximum vacancy content was found to result in highly ordered monoclinic structures where manganese occupies two of the $\sqrt{3} \times \sqrt{3}$ superlattices on the transition metal layers while the third was randomly filled with nickel, lithium and vacancies. The resulting ordering predicted by a Monte Carlo simulation was consistent with the sharp ordering peaks seen in the XRD patterns. The vacancy concentrations were confirmed by Rietveld refinement, density measurements and redox titration; all of which were in good agreement. The role of the vacancies during electrochemical cycling, if there is one, remains unclear. The material with the greatest possible vacancy concentration, $Li[Ni_{1/6}\square_{1/6}Mn_{2/3}]O_2$, showed electrochemical behavior consistent with lithium-rich layered materials, namely high irreversible capacity associated with the high voltage plateau and voltage fade associated with conversion to spinel. However, this material was not, in fact, lithium-rich given that the vacancies result in lithium occupying only 50 % of metal sites. The vacancy results also demonstrated that there were roughly 1 % vacancies in a stoichiometric lithium-rich material lying along the line from Li_2MnO_3 and $LiNi_{0.5}Mn_{0.5}O_2$. The Monte Carlo simulation suggested that this allows Ni^{3+} to substitute for Mn^{4+} on two of the superlattices. This has never been recognized before and a complete understanding of the starting material is crucial to fully understand the complex electrochemical behavior of the lithium-rich positive electrode materials. The exact shape of the other side of the bump (to the left of $Li[Ni_{1/6}\square_{1/6}Mn_{2/3}]O_2$ in the Gibbs triangles) has not been determined. It is also unclear as to why the bump would be so sharp on both sides and this is worthwhile for further study.

Chapter 9
Materials Near the Layered Boundary

9.1 Motivation for Studying $LiNi_{0.5}Mn_{0.5}O_2$

As previously discussed, $LiNi_{0.5}Mn_{0.5}O_4$ lies near the upper boundary of the layered region. Figure 9.1 shows the phase boundaries of the layered region when heated in air to either 800 or 900 °C as determined in previous chapters. The effect of synthesis atmosphere and cooling rate will be studied for two compositions in the current chapter, labeled A_9 and B_9. The single-phase layered boundary moves downward in the Gibbs triangle when samples are cooled more slowly such that sample A_9, which is single-phase if quenched, lies in the layered–layered two phase coexistence region when cooled more slowly. A higher oxygen partial pressure also lowers the upper layered boundary in the Gibbs triangle, since more oxygen favors the spinel structures over the relatively oxygen-poor layered structures. Thus, for lower oxygen partial pressures, sample A_9 will lie very close to the layered boundary when regular cooled. The main objective of this chapter is to find conditions that produce a sample showing the first signs of layered–layered phase separation in order to study the consequences of this on the performance of the electrode material.

In the literature, there is considerable debate over whether or not lithium-rich layered materials form solid solutions [2, 3, 46] or layered–layered nano-composites [43, 44]. In the combinatorial study, the compositions where one can expect layered–layered materials when synthesized in oxygen were determined. The fact that these materials can transform dramatically during slow cooling has been known for quite some time [11, 54]. Kang and Amine [11] observed that a single-phase sample of $Li_{1.17}Ni_{0.25}Mn_{0.58}O_2$ had a first cycle capacity of 175 mAh/g when quenched and showed phase separation in the X-ray diffraction (XRD) pattern when slow cooled giving a first charge capacity as low as 55 mAh/g and never exceeding 85 mAh/g. It is difficult to identify the phases present based on the XRD shown in Ref. [11]. However, based on the phase diagrams from Chaps. 5–7, their sample was made up of a monoclinic phase near M, a rocksalt phase and/or a layered phase near

Data in this chapter are included in Ref. [92], published in the Journal of the Electrochemical Society.

E. McCalla, *Consequences of Combinatorial Studies of Positive Electrodes for Li-ion Batteries*, Springer Theses, DOI 10.1007/978-3-319-05849-8_9,

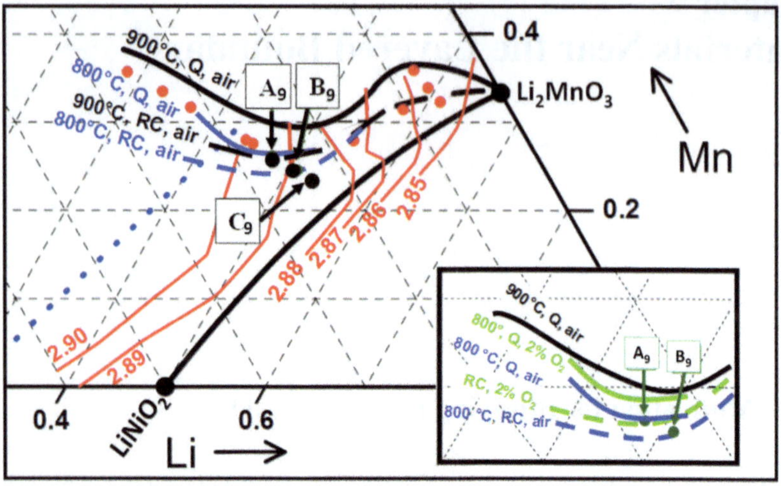

Fig. 9.1 Partial phase diagram with contours for the *a* lattice parameter showing the boundaries of the single-phase layered region. The lower boundary is shown connecting $LiNiO_2$ to Li_2MnO_3 (no attempt was made to determine how this boundary changes with synthesis conditions, since the current study focuses on compositions near the top of the layered region). The insert shows approximate upper boundaries of the layered region consistent with the current study. Points A_9 and B_9 were determined using elemental analysis and are referred to throughout the text. The *red points* are discussed in Chap. 7. The *blue dotted line* is the cubic to layered phase transition

N. Nonetheless, the fact that such phase separation gives rise to such poor electrochemistry has been seen before, but the cause of this loss in capacity remains unclear and is the subject of this chapter.

Since there is a great deal of differing data in the literature for materials reported as $LiNi_{0.5}Mn_{0.5}O_2$, this chapter will also help clarify why there has been such a wide spread in results. Ohzuku et al. first reported on this material and showed reversible cycling capacity of 150 mAh/g up to 4.3 V [86]. Lu et al. [2] demonstrated that quenched samples were cycled with a capacity of 140 mAh/g reversibly up to 4.4 V and 190 mAh/g up to 4.8 V with only 25 mAh/g irreversible capacity (IRC). These samples were made by coprecipitation with LiOH such that a small amount of excess lithium may have been present. The effect of excess lithium will be examined in detail here in the discussion of a set of $Li_{1+x}[Ni_{0.5}Mn_{0.5}]_{1-x}O_2$ samples. An extreme example of this was obtained by Ohzuku et al. [52] who reported a sample of $LiNi_{0.5}Mn_{0.5}O_2$ made with 25 % excess lithium that cycled 200 mAh/g stably up to 5.0 V with a larger IRC of 50 mAh/g. By contrast, Lin et al. [53] made $LiNi_{0.5}Mn_{0.5}O_2$ without any excess lithium. The quenched sample had a first cycle discharge capacity of 175 mAh/g over the range 2.5–4.5 V and faded to 125 mAh/g after 50 cycles while a regular cooled sample showed stable cycling of 125 mAh/g over 30 cycles. Also, the XRD patterns of the samples from Lin et al. showed some ordering peaks in the range 20–30° in the regular cooled sample only. This is consistent with phase separation where one of the phases has ordering on the transition metal (TM) layer.

However, the XRD patterns in the Lin paper are not sufficiently detailed to determine whether or not peak broadening was present in the regular cooled sample. Part of the current study therefore deals with reproducing these data in order to determine whether peak broadening is associated with this decrease in first cycle capacity.

It was also necessary to identify the features in the XRD patterns which are most sensitive to phase separation and then search for them carefully. In Chap. 4, in the Li–Co–Mn–O system, peak broadening at high angles (>60°) in the XRD patterns was found to first indicate the formation of layered–layered composites and this corresponded to synthesis conditions where Wen et al. [22] found nano-composites. Upon slower cooling of the combinatorial samples described in Chap. 4, the broad peaks eventually resolved into two separate peaks corresponding to each layered phase such that the formation of nano-composites can be viewed as an incomplete phase separation into a layered–layered two-phase structure. To our knowledge, this was the first time that features in XRD patterns were correlated to nano-scale phase separation. Similar careful analysis of XRD patterns will be applied to samples in the Li–Mn–Ni–O system here. Therefore, the condition used to identify the presence of nano-composites in the XRD pattern was the broadening of the high angle peaks. However, to distinguish this from microstrain peak broadening, a necessary condition to identify the formation of a nano-composite is that clear phase separation must be seen in the XRD if conditions are exaggerated (e.g., if cooling rates are slowed down further as done in Chap. 4 or if oxygen partial pressures are increased as used here).

9.2 Experimental Design

In order to obtain a sample showing these first signs of phase separation, the boundaries of the single-phase region were varied by changing the oxygen partial pressure during heating. Figure 9.1 shows the upper layered boundary under various conditions. The boundaries for samples heated in air were presented in previous chapters while the approximate boundaries for 2 % oxygen are based on the current study and so should only be considered accurate near compositions A_9 and B_9. The boundary moves upward as oxygen partial pressure decreases. Therefore, a sample made in pure oxygen at composition A_9 will be multiphase if quenched while the same composition in 2 % oxygen will be single phase. As will be shown in the results section, the regular cooled sample made at composition A_9 in 2 % oxygen showed the signs of being a layered–layered nano-composite. All samples made by the one-pot synthesis method at composition A_9 were heated at the rate of 5 °C/min up to 800 °C and held there for 5 h. The atmospheres used during heating were either air, or a flow of gas with either 2, 5.5, or 100 % oxygen content with the remainder being either nitrogen or argon gas. The samples were then either quenched or regular cooled. Samples at the composition B_9 were made by Aaron Rowe and Eric McCalla using the tank reactor method. Samples were heated at 900 °C for 12 h before either quenching or regular cooling.

Electrochemical tests were carried out on some samples as described in Sect. 2.5. Two identical cells were made for each sample. All cycling was performed at a specific current of 10 mA/g. For samples A_9 and B_9, made by regular cooling after heating in the 2 % oxygen mixture, inductively coupled plasma (ICP) was used as elemental analysis to determine the metal molar fractions of the final products (these values were used in generating Fig. 9.1). The XRD patterns from all samples were measured in either a JD-2000 diffractometer or a Siemens D-5000 diffractometer. The resulting scans were then analyzed using Rietveld refinement.

A series of samples $Li_{1+x}[Ni_{0.5}Mn_{0.5}]_{1-x}O_2$ was also synthesized at 900 °C in air with regular cooling (by Jing Li who performed all measurements on these samples). This composition line was previously studied by Myung et al. [64] over the range $0 \leq x \leq 0.06$ and capacity was found to increase with x when cycled up to 4.6 V. Here, the wider composition range $0 \leq x \leq 0.24$ was studied and the upper cutoff was 4.4 V in order to avoid the high voltage plateau [2]. Samples A_9, B_9 and C_9 in Fig. 9.1 have nominal compositions given by $x = 0$, 0.04, and 0.08, respectively. Samples with $x = 0.12$, 0.16, 0.20, and 0.24 were also prepared.

Finally, to better understand the consequences of the layered–layered phase separation on the structure, a Monte Carlo simulation of a material of composition A_9 in Fig. 9.1 was performed. The details of this simulation are given in Sect. 2.12. The temperature scale was set by the parameter ,β_T and previous results suggest that $\beta_T = 1$ corresponds to a temperature near and below 800 °C, such that the value of $\beta_T = 2$ used here is near 250 °C, and is therefore well below the critical temperature above which solid solutions are favored due to a higher entropy. Two compositions were simulated: $Li_{0.9}Ni_{0.1}[Ni_{0.4}Mn_{0.5}Li_{0.1}]O_2$ (stoichiometric $LiNi_{0.5}Mn_{0.5}O_2$ with 10 % nickel on the lithium layer, consistent with Ref. [2]) and $Li[Ni_{0.4}Mn_{0.5}Li_{0.1}]O_2$ simulated in order to determine the role played by nickel on the lithium layer. The fact that there was nickel on the lithium layer in the first simulation meant that the model required out-of-plane interactions. Therefore, two layers were simulated: one lithium layer and one TM layer. It was assumed that these two layers stack alternately and the only out-of-plane interactions included in the calculations were for nearest neighbors, such that every atom had six in-plane and six out-of-plane nearest neighbor interactions. Periodic boundary conditions were used.

9.3 Structural Results

The results of the elemental analysis for composition A_9 were $Li_{0.48}Ni_{0.26}Mn_{0.26}$ and that point was positioned accordingly in Fig. 9.1. This composition suggests that the material, if single phase, would be $Li_{0.96}Ni_{0.52}Mn_{0.52}O_2$ assuming no metal/oxygen vacancies and can therefore be considered a slightly lithium-poor layered material. Figure 9.2 shows XRD scans of samples at composition A_9 made under various conditions. The scan of the sample made in pure oxygen and regular cooled shows three phases (three peaks visible near 44°) consistent with the R, N, and M phases discussed at length in previous chapters. Since broadening of the Bragg peak near

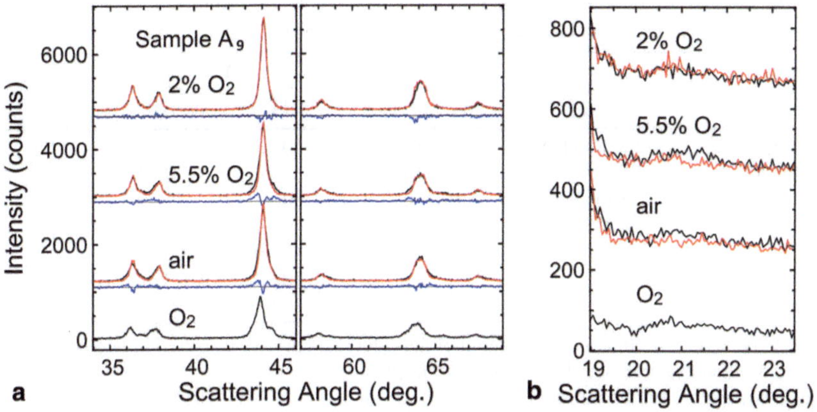

Fig. 9.2 X-ray diffraction (XRD) scans of samples with composition A_9 heated at 800 °C. *Black lines* represent samples that were regular cooled, while *red lines* are for quenched samples. The difference plots in *blue* in the *left panel* represent the quenched scan subtracted from the regular cooled scan

44° is a clear indication of phase separation, the scans taken after synthesis in air and 5.5 % oxygen show clear signs that phase separation occurs during regular cooling. This is visible in the difference plots (calculated as the quenched pattern subtracted from the regular cooled one). The shape of the difference plots near 44° is consistent with broadening of the peak on both sides. By contrast, the sample made in 2 % oxygen showed no such peak broadening near 44°. The right panel of Fig. 9.2 shows the region where superlattice peaks are expected if there is ordering on the TM layers. There are no peaks visible here for all quenched samples consistent with the disordered TM layer expected for $LiNi_{0.5}Mn_{0.5}O_2$. However, the regular cooled samples in 5.5 % oxygen, air, and pure oxygen show small broad peaks near 21°. This is consistent with phase separation wherein one of the new phases is a layered material with ordering on the TM layer as seen in the M-layered phase in Chap. 6.

To more carefully quantify the changes taking place during cooling, the scans were fit using Rietveld refinement assuming a single hexagonal layered structure (R-3m space group). Table 9.1 shows the results of the refinement. For all three quenched samples made at composition A_9, the lattice parameters are consistent with previously presented contour plots (the *a* lattice parameter is shown in Fig. 9.1, the *c* values can be found in Chap. 7). Table 9.1 also shows the widths of the calculated peaks at 44 and 65°. The extreme broadening seen in the 5.5 % oxygen and air samples suggests once again that these materials phase separated during regular cooling. By contrast, the 2 % oxygen regular cooled sample showed no peak broadening at 44° and moderate broadening of the higher angle peak, accompanied with a decrease in the *c* lattice parameter suggesting a broadening to the high angle side of peaks. The same behavior was seen in regular cooled samples of Li–Co–Mn–O under conditions where layered–layered nano-composites were seen (Chap. 4)—a broadening of high angle peaks only and clear phase separation if conditions are exaggerated (i.e., the samples

Table 9.1 X-ray diifraction (XRD) Rietveld refinement results for samples of composition A_9 fit as single-phase layered. *Q* denotes quenched while *RC* represents regular cooled samples

Synthesis	a(Å)	c(Å)	FWHM[a]		Ni_{Li}(%)[c]
			44°	65°	
2 % O_2/Q	2.8986 (11)	14.312 (4)	0.43	0.43	13
5.5 % O_2/Q	2.8983 (10)	14.315 (4)	0.40	0.44	12
Air/Q	2.8968 (14)	14.310 (3)	0.43	0.49	13
2 % O_2/RC	2.8986 (11)	14.293 (4)	0.44	0.52	11
5.5 % O_2/RC	[b]	[b]	0.50	0.62	[b]
air/RC	[b]	[b]	0.52	0.63	[b]
100 % O_2/RC	[b]	[b]	0.67	0.75	[b]

[a] FWHM is the full width at half maximum of the calculated peaks, measured in degrees
[b] Values omitted as these samples contained multiple phases
[c] Nickel occupation on the lithium layer

prepared under higher oxygen partial pressures showed clear phase separation here. The sample A_9 heated in 2 % oxygen therefore shows the signs of forming a layered–layered composite on short length scales when regular cooled. The sample made in air and quenched also shows broadening at the higher angle peak only (though less so with a full width at high maximum (FWHM) value of 0.49) suggesting that in air, the boundary for quenched samples is quite close to point A_9. This again suggests that the phase boundary was lower in the triangle as the oxygen partial pressure was increased, such that the sample made in air may have already begun to phase separate on short length scales even when quenched. The Rietveld refinement results also show that the fraction of nickel on the lithium layer is slightly above 10 % which is consistent with previous studies [2]. It would be of high interest to study such samples with methods sensitive to short-range ordering such as high resolution transmission electron microscopy in order to distinguish between phase separation over short distances and phase separation between two phases very close to each other on the phase diagram. In either case, the XRD patterns indicate that the layered samples at A9 all phase separated during regular cooling, with the sample made in 2 % O_2 showing the smallest signs of multiple phases.

For comparison, a regular cooled sample synthesized in 2 % oxygen at composition B_9 was found to have the composition $Li_{1.02}Mn_{0.50}Ni_{0.48}O_2$ according to elemental analysis and Rietveld refinement yielded $a = 2.8891$ Å and $c = 14.296$ Å, again consistent with the contour plots shown in Fig. 9.1 and Chap. 7. The XRD patterns for this sample showed no peak broadening after regular cooling and as such indicated that the sample stayed single phase. The boundaries in Fig. 9.1 were drawn to be consistent with all XRD results discussed in this thesis. The 2 % oxygen boundary is therefore approximate and should only be considered valid near the compositions A_9 and B_9, since the rest of the phase diagram has not been studied extensively under these conditions. Similarly, since it was difficult to precisely determine the position of the boundaries for the 800 °C quenched samples, it was drawn going through point A_9 given that a small amount of peak broadening was seen in the high angle peaks in this sample.

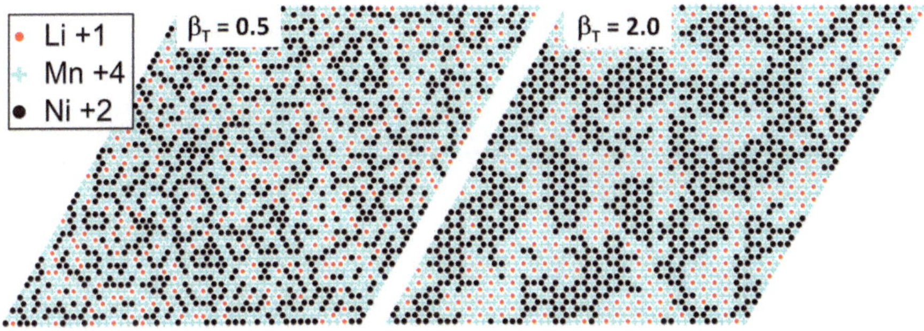

Fig. 9.3 Monte Carlo simulation results for the transition metal layer with composition $Ni_{0.4}Mn_{0.5}Li_{0.1}$ and no nickel on the lithium layer

9.4 Monte Carlo Simulation Results

Figure 9.3 shows the results of the Monte Carlo simulation for the composition $Li[Ni_{0.4}Mn_{0.5}Li_{0.1}]O_2$ (i.e., no nickel present on the lithium layer). The results at $\beta_T = 0.5$ (very high temperature) showed a solid solution where sites were randomly occupied except for lithium coordinating around Mn on the TM layer as has been well documented before. Upon cooling to $\beta_T = 2.0$ ($T \propto 1/\beta_T$), phase separation was evident with clusters of pure $LiMn_2$ separated by regions with nickel and manganese only of approximate composition $Ni_{0.5}Mn_{0.5}$. This phase separation is promoted heavily in the literature [43, 44]. Figure 9.4, however, shows that phase separation is strongly hindered when nickel is present on the lithium layer. The $LiMn_2$ clusters stayed much smaller and also contained some nickel. This Ni in the $LiMn_2$ clusters is consistent with the phase diagram where coexistence includes a structure close to the M-layered material which contains a small amount of nickel on the TM layer. Allowing the simulation to run 10 times longer (labeled annealed) did promote phase separation somewhat though domains were still smaller than those seen in Fig. 9.3 and there was still disorder in the domains.

Another significant feature seen in the Monte Carlo simulation results in Fig. 9.4 was that in the regions of disordered $LiMn_2$ on the TM layer, there was no nickel on the corresponding region in the lithium layer. This implies that nickel in the lithium layer is clustering to the regions of $Ni_{0.5}Mn_{0.5}$ on the TM layers. Once again, this feature is expected from the phase diagram given that the N phase has roughly 30 % nickel on the lithium layer while the M phase only contains approximately 2 % such that M–N phase separation would result in clustering of nickel on the lithium layers. Here, the endpoints lie along the M–N segment of the boundary such that some nickel clustering on the lithium layer is expected. These results suggest that the ordering of $LiMn_2$ on the TM layer may drag lithium on the lithium layer. This effect is consistent with the tendency of lithium to coordinate around manganese though this coordination is now between two neighboring layers. The consequence of lithium coordinating to manganese is that nickel must cluster on the lithium layer. This may create a severe hindrance to the diffusion of lithium in the lithium layer.

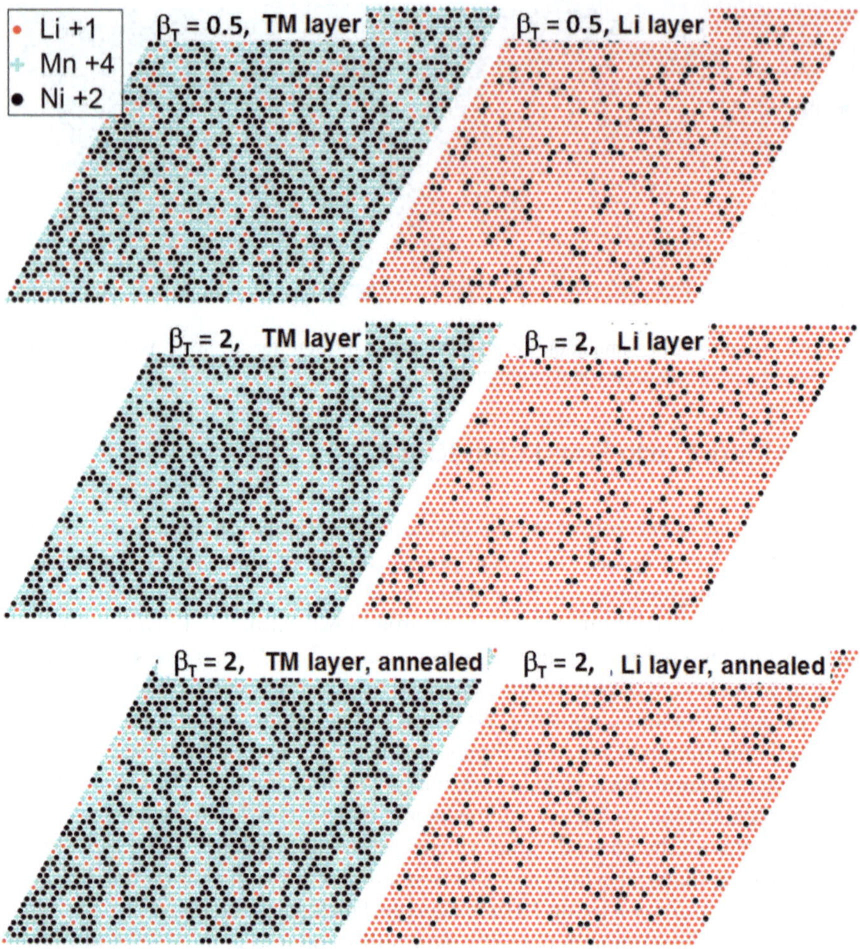

Fig. 9.4 Monte Carlo simulation results of sample $LiNi_{0.5}Mn_{0.5}O_2$ with 10 % nickel on the lithium layer. The transition metal (TM) layer had composition $Ni_{0.4}Mn_{0.5}Li_{0.1}$. Annealed indicates 10 times as many Monte Carlo cycles

9.5 Electrochemical Measurements

Figure 9.5 shows the capacity versus cycle number for cells made from materials made under either 2 % oxygen or in air. The quenched and regular cooled materials at composition A_9 made in air are consistent with those seen by Lin et al. in Ref. [53] with the quenched sample here showing a capacity of 150 mAh/g fading rapidly with cycling up to 4.4 V and the regular cooled sample cycling about 100 mAh/g. No efforts were made here to determine the changes taking place during cycling that give rise to this capacity fade and this question remains a significant one. The higher

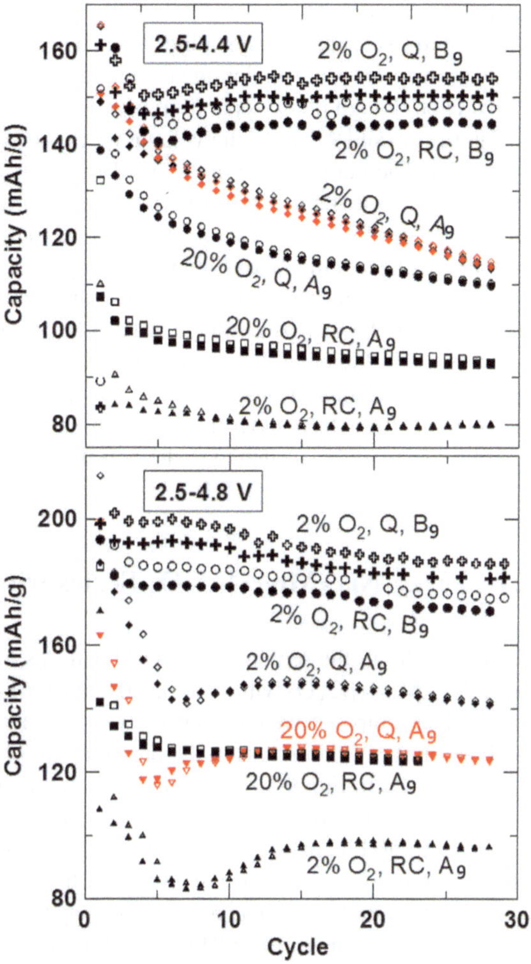

Fig. 9.5 Capacity versus cycle number for samples A_9 heated to 800 °C and B_9 heated to 900 °C cycled at 10 mA/g. *Q* is for quenched samples, *RC* is for regular cooled, while 20 % oxygen is for samples heated in air. In the *top panel*, twin cells are included for the 2 % oxygen sample when quenched in order to show the reproducibility

capacities (190 mAh/g up to 4.8 V and 150 mAh/g up to 4.4 V) seen for samples at composition B_9 heated to 900 °C and either quenched or regular cooled confirm the importance of staying within the single-phase region during cooling. These values are also consistent with those reported elsewhere [2, 86] for $LiNi_{0.5}Mn_{0.5}O_2$ samples made with a small amount of excess lithium.

Of higher interest for the current study was the behavior of the A_9 samples made in 2 % oxygen. When the electrode material was quenched, the cell capacities for samples heated in 2 % oxygen were slightly higher than those made in air and show the same capacity fade with cycling. However, upon regular cooling, the capacity dropped dramatically to be well below that of the material made in air, for both upper voltage cutoff limits of 4.4 and 4.8 V. This large drop in capacity took place even though the 2 % oxygen samples showed the least sign of phase separation based on the XRD. This may be due to the clustering of nickel on the lithium layer on a single crystallite blocking lithium diffusion paths as would occur in a nano-scale composite material. Then, upon completing the phase separation, the materials made in air would have separate crystallites of each phase such that the more ordered phase would then cycle nicely explaining the higher capacities obtained in air when regular cooled even though phase separation was more evident in the XRD.

Figures 9.6 and 9.7 show the voltage vs. capacity and dQ/dV vs. voltage plots for these cells. The main feature in the dQ/dV plots was a peak near 3.75 V with some smaller peaks near 4.5 V corresponding to a small high voltage plateau in the capacity versus voltage plots. The most relevant features in the cells made from regular cooled A_9 materials heated in 2 % oxygen were a large irreversible capacity when cycled up to 4.8 V and a significant impedance growth in both regular cooled materials. This impedance can be seen as a voltage difference between charge and discharge, which was noticeably larger in the regular cooled samples as compared to the quenched samples. This change was also seen in the samples made in air, though it was less severe, while the cells made from sample B_9 showed no such impedance growth such that this can again be attributed to the presence of multiple phases appearing during cooling.

9.6 $Li_{1+x}[Ni_{0.5}Mn_{0.5}]_{1-x}O_2$ Series with $0 \le x \le 0.24$

Figure 9.8 shows the region of the XRD patterns near 44° for the composition line $Li_{1+x}[Ni_{0.5}Mn_{0.5}]_{1-x}O_2$ with $x = 0.0$, 0.04, and 0.08 made at 900 °C in air and regular cooled. The samples have not been analyzed by elemental analysis, but based on the small amount of lithium loss in the previous samples, these samples lie near points A_9, B_9, and C_9 in Fig. 9.1. The sample with $x = 0.00$ (point A_9) shows evidence for the presence of a second phase, consistent with the XRD pattern expected for sample A_9 heated to 900 °C in air and regular cooled in Fig. 9.1. The samples with $x = 0.04$ and 0.08 do not show evidence of a second phase, although the peak width increases with x. The crystallite size and microstrain may therefore depend on the composition, x. Table 9.2 summarizes the XRD results for the $Li_{1+x}[Ni_{0.5}Mn_{0.5}]_{1-x}O_2$ series with $0 \le x \le 0.24$. The XRD patterns therefore confirm that changing the value of x in this manner moved samples from outside the single-phase region (sample A_9, $x = 0.0$) to within the single-phase region (samples B_9 and C_9), and then outside the single-phase region, where excess Li_2CO_3 was observed in the XRD patterns ($x = 0.20$ and 0.24). Studies of these samples also allow the determination of the impact of the layered–layered phase separation as all samples were made under identical conditions.

Figure 9.9 shows the reversible specific capacity of the $Li_{1+x}[Ni_{0.5}Mn_{0.5}]_{1-x}O_2$ samples during the first and fourth discharges. This figure shows that samples should be prepared as single-phase materials in order to deliver the highest capacity and that phase separation into layered–layered composites ($x = 0$ in Fig. 9.9) should be avoided. Figure 9.9 also shows that samples closest to the single-phase boundary have the highest capacity when cycled to 4.4 V. This illustrates the importance of a complete understanding of the phase diagram in order to avoid Ni^{3+} in the starting material and to produce materials with the best electrochemical performance.

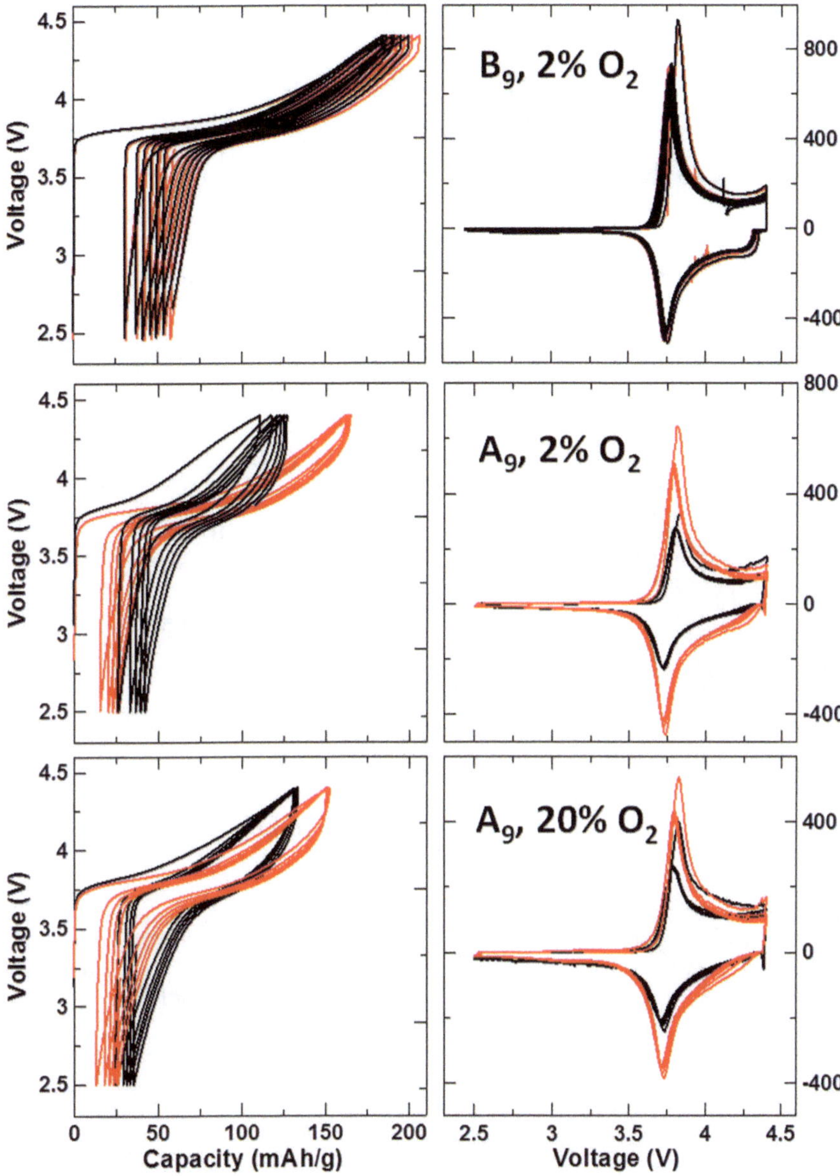

Fig. 9.6 Voltage curves and dQ/dV for samples A_9 and B_9 cycled up to 4.4 V. The *red lines* are for quenched samples, while the *black lines* are for regular cooled

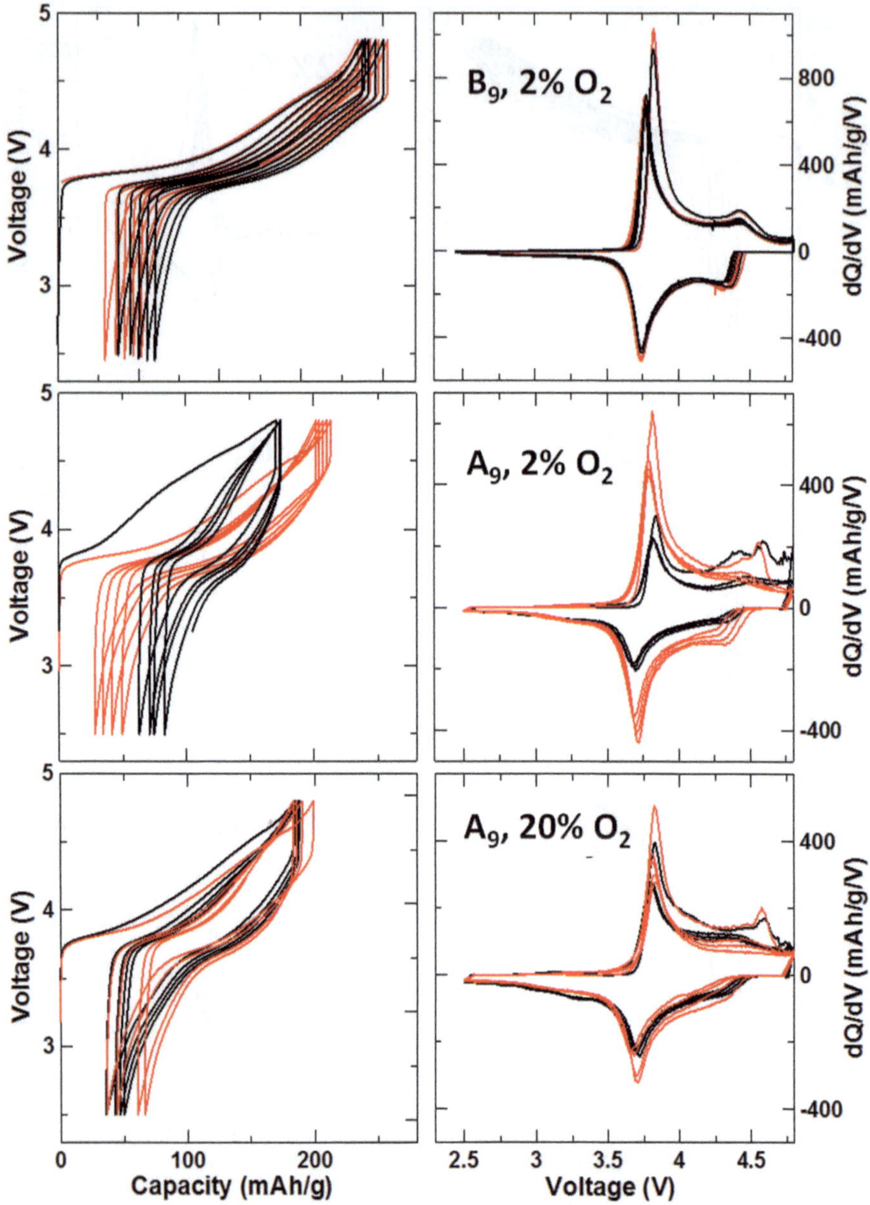

Fig. 9.7 Voltage curves and dQ/dV for samples A_9 and B_9 cycled up to 4.8 V. The *red lines* are for quenched samples, while the *black lines* are for regular cooled

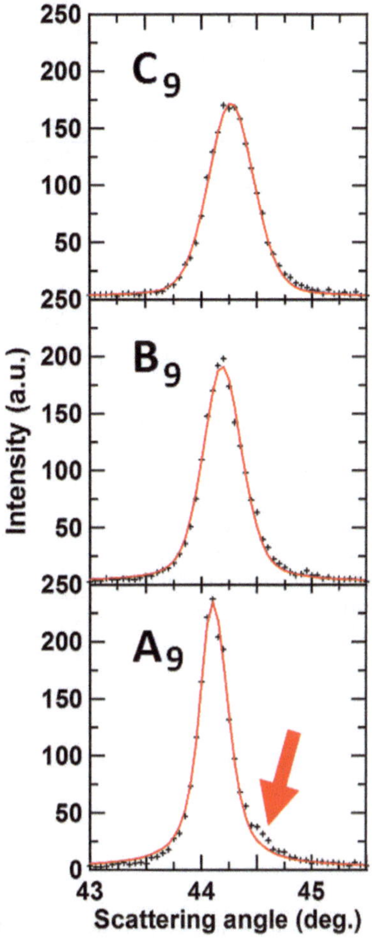

Fig. 9.8 Partial X-ray diffraction (XRD) patterns for samples A_9 ($x = 0.00$), B_9 ($x = 0.04$) and C_9 ($x = 0.08$) obtained as part of the composition series $Li_{1+x}[Ni_{0.5}Mn_{0.5}]_{1-x}O_2$. The *arrow* indicates the presence of a second peak in sample A_9. The *red lines* are single-phase layered fits

9.7 Conclusions Regarding Layered–Layered Nano-Composites

The structural changes taking place during regular cooling of materials with metallic compositions near that of $LiNi_{0.5}Mn_{0.5}O_2$ were studied after heating under various oxygen partial pressures and were related to features in the phase diagram determined previously. Materials made with 5 % excess lithium (at point B_9 in Fig. 9.1) were found to stay single phase even when the regular cooling rate of 5–10 °C/min was used. These single-phase layered materials showed good electrochemical performance with a reversible capacity of 180 mAh/g for the quenched sample and 170 mAh/g for the regular cooled sample when cycled up to 4.8 V. By contrast, under the same cycling conditions, samples without excess lithium (at point A_9 in Fig. 9.1) made in 2 % oxygen had a capacity of 140 mAh/g when quenched and only 90 mAh/g when regular cooled. The changes in the XRD pattern for this A_9 sample

Table 9.2 X-ray diffraction (XRD) Rietveld refinement results for samples along the composition line $Li_{1+x}[Ni_{0.5}Mn_{0.5}]_{1-x}O_2$ with $0 \leq x \leq 0.24$. The capacity is the average first cycle discharge capacity for two twin cells cycled over the range 2.5–4.4 V at 10 mA/g

Sample	x	a ($\pm$0.0004 Å)	c ($\pm$0.002 Å)	Ni_{Li} (%)	Single phase	Capacity (mAh/g)
$LiNi_{0.5}Mn_{0.5}O_2$	0.00	2.8944	14.311	10.0	No	129.7(7)
$Li_{1.04}Ni_{0.48}Mn_{0.48}O_2$	0.04	2.8890	14.307	9.0	Yes	160.3(5)
$Li_{1.08}Ni_{0.46}Mn_{0.46}O_2$	0.08	2.8816	14.293	7.4	Yes	149.9(7)
$Li_{1.12}Ni_{0.44}Mn_{0.44}O_2$	0.12	2.8765	14.279	5.0	Yes	142(1)
$Li_{1.16}Ni_{0.42}Mn_{0.42}O_2$	0.16	2.8704	14.255	3.7	Yes	134.3(5)
$Li_{1.20}Ni_{0.4}Mn_{0.4}O_2$	0.20	2.8678	14.250	2.2	No[a]	129.7(5)
$Li_{1.24}Ni_{0.38}Mn_{0.38}O_2$	0.24	2.8679	14.242	0.7	No[a]	121(2)

[a] This coexistence is between the layered material and unreacted Li_2CO_3

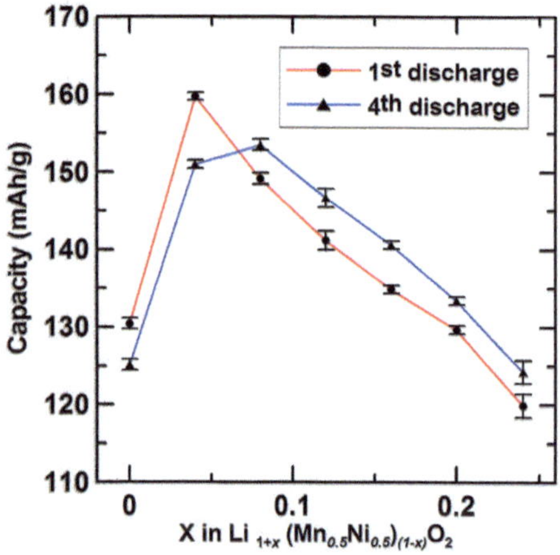

Fig. 9.9 Discharge capacity as a function of nominal composition x in $Li_{1+x}[Ni_{0.5}Mn_{0.5}]_{1-x}O_2$. Cells were cycled over the range 2.5–4.4 V at 10 mA/g. The *error bars* represent the variation between two twin cells

were small with peak broadening only seen at high angle, consistent with the first sign of phase separation into layered–layered nano-composites. The dramatic loss in capacity seen in the sample made in 2 % oxygen at point A_9 can be attributed to nickel clustering on the lithium layer such that lithium islands form, many of which would be surrounded by the clustered nickel. This behavior was expected from the phase diagram where phase separation into a nickel rich and nickel poor phase was demonstrated in Chap. 6, and this clustering of nickel was also seen in a Monte Carlo simulation performed in this chapter.

Samples made in air, which showed greater signs of phase separation, had a smaller decrease in capacity when regular cooled, consistent with phase separation reaching completion, such that whole crystallites of each phase exist and the more ordered layered material then delithiates more easily. Samples of $Li_{1+x}[Ni_{0.5}Mn_{0.5}]_{1-x}O_2$ which were made in air at 900 °C showed the highest capacity when they were single phase near the phase boundary. This work therefore shows the importance of precisely knowing where the samples lie in the phase diagram and carefully examining XRD peak shapes in order to detect the smallest signs of phase separation. It would now be of high value to collect short-range information on materials near the layered boundary from techniques such as transmission electron microscopy in order to more carefully determine the compositions where layered–layered composites form on the nano-scale and to correlate this to the results discussed here. It is important to realize that such electron microscopy experiments would need to be done with extreme care in order to avoid situations where apparent phase separation might be caused by electron beam heating.

Chapter 10
Conclusions and Future Works

10.1 The Li–Co–Mn–Ni–O Pseudo-Quaternary System

Figure 10.1 shows the Li–Co–Mn–Ni–O pseudo-quaternary system with the Li–Co–Mn–O and Li–Mn–Ni–O faces shown, as determined with combinatorial samples quenched from 800 °C. Some approximations were made to join the two faces since the Li–Co–Mn–O face was synthesized in air while the nickel containing samples were made in oxygen. Nonetheless, the two faces join quite well. The pyramid strongly suggests that both single-phase regions of importance for battery materials, spinel and layered, extend into the pyramid and form relatively large 3D shapes. On the Li–Co–Mn–O system, the layered region is restricted to a single line showing that cobalt is always synthesized in the 3+ state as it is in $LiCoO_2$. By contrast, nickel can be in the 2+ state as in NiO rocksalt or the 3+ state as in layered $LiNiO_2$ such that a much larger and more complex layered region exists on the Li–Mn–Ni–O face. The spinel-layered coexistence region is also simpler in the Li–Co–Mn–O triangle with all tie-lines connecting to either the cobalt spinel, Co_3O_4, or the manganese layered material, Li_2MnO_3, while in the Li–Mn–Ni–O system there are 2 three-phase regions. The differences between the roles of cobalt and nickel should prove significant in upcoming combinatorial work in the Li–Co–Mn–Ni–O pseudo-quaternary system that is of extreme interest for battery materials as it includes commercial materials such as $Li[Ni_{1/3}Mn_{1/3}Co_{1/3}]O_2$ [93] as well as promising spinel-layered core-shell materials [88] and lithium-rich layered materials [3].

One of the main findings in this work is that the boundaries of the single-phase region move dramatically with synthesis conditions. Most importantly, the layered single-phase regions become smaller when cooled more slowly, and this occurred in both faces of the pyramid. The consequences of this were discussed in detail throughout the thesis and will be summarized here by showing how this work helps resolve a number of points of confusion in the literature mentioned in Chap. 1. These were listed with roman numerals and will be repeated and addressed here with the same numerals in the next section.

E. McCalla, *Consequences of Combinatorial Studies of Positive Electrodes for Li-ion Batteries*, Springer Theses, DOI 10.1007/978-3-319-05849-8_10,

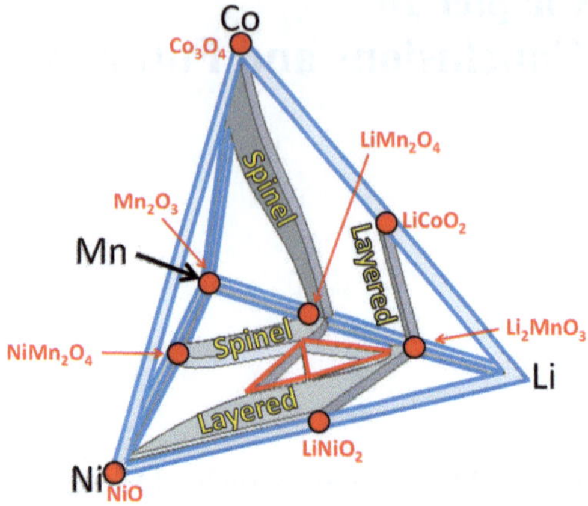

Fig. 10.1 The Li–Co–Mn–Ni–O pseudo-quaternary system for samples heated to 800 °C and quenched with single-phase regions identified. The *red lines* indicate boundaries of three-phase regions while the *blue lines* denote the axes

10.2 Resolving Points of Confusion

I In the Li–Co–Mn–O system, Kim et al. [18] stated that the composition line joining $LiCoO_2$ to Li_2MnO_3 forms a solid-solution while Bareño et al. [23] determined that samples near the center of the line phase separate into layered–layered composites on the 2–10 nm length scale. The combinatorial work described in Chap. 4 demonstrated that the line was a solid solution if quenched from at least 800 °C, implying that at these temperatures the high entropy of the solid solutions result in the lowest free energies. However, upon extreme slow cooling, samples clearly phase separated near the center of the composition line. This showed that internal energy plays a greater role below 800 °C and these results were confirmed with a Monte Carlo simulation. At intermediate cooling rates, X-ray diffraction (XRD) peaks broadened at compositions where phase separation was seen upon slow cooling. This broadening was attributed to the formation of layered–layered nano-composites as seen by Bareño. Thus, the center of the solid-solution phase separates into nano-composites during regular cooling.

II Bareño et al. [23] claimed that the phase separation along the $LiCoO_2$ – Li_2MnO_3 line involved the formation of $LiCoO_2$ (Co on the transition metal (TM) layers) and Li_2MnO_3 ($Li_{1/3}Mn_{2/3}$ on the transition metal (TM) layers) domains. The combinatorial work showed that the phase separation does not occur over the whole composition line. Even at the slow cooling rate of 1 °C/min the two domains were roughly 80 % Co, 20 % $Li_{1/3}Mn_{2/3}$ and 20 % Co, 80 % $Li_{1/3}Mn_{2/3}$. This result suggests that although careful TEM studies can be useful in finding nano-composites, another method is required to determine the compositions of the domains present. It is also important to note that the phase separation can be avoided entirely by quenching or making materials at compositions near the end-members of the composition line.

III Lithium loss during synthesis is either viewed as Li_2O evaporation [30] or the formation of lithium peroxide vapour [31]. In the process of optimizing the combinatorial synthesis method, it was found that lithium loss occurred primarily by way of the formation of lithium peroxide or, at the very least, involved Li_2O reacting with oxygen. Samples heated in oxygen-free atmospheres showed solid lithium oxide in the XRD while samples heated in oxygen-containing atmospheres did not. This implies that LiO_2 does not evaporate under the conditions used to make electrode materials, but rather it must react with oxygen to leave the sample.

IV A number of researchers have found a rocksalt contaminant in either spinel or layered-spinel materials in the Li–Mn–Ni–O system [34, 37, 38]. The composition and structure of this contaminant was under debate prior to the current thesis. The results of the combinatorial study show that the contaminant, when slow cooled in oxygen, had metallic composition $Li_{0.22}Mn_{0.16}Ni_{0.62}$ which lies well inside the Gibbs triangle and not on the Li–Ni line as previously believed. The structure of this material is a cubic rocksalt with a 2 x 2 x 2 superlattice with ordering of manganese on one site and lithium, nickel, and vacancies on another.

V The $LiNi_{0.5}Mn_{1.5}O_4$ spinel material phase separates in air forming the rocksalt material already mentioned and a spinel lying higher in the Gibbs triangle [34]. However, $LiNi_{0.5}Mn_{0.5}O_4$ may also accommodate oxygen vacancies [39, 40]. It is difficult to distinguish these two processes in air because the sample is two-phase and both the phase separation and the formation of oxygen vacancies involve the loss of oxygen and an increase in lattice parameter. However, in oxygen, this composition is single-phase and the lattice parameter contour plots in Chap. 5 clearly show a larger lattice parameter for the quenched sample. This shows that the Li–Mn–Ni–O spinel materials do sustain oxygen vacancies at high temperature, whereas the Li–Mn–O spinels do not as the lattice parameters do not change dramatically here.

VI The lithium-rich layered line from $LiNi_{0.5}Mn_{0.5}O_2$ to Li_2MnO_3 was considered to be either solid solutions [2] or layered–layered nano-composites [43] in the literature. Once again, the results from this project showed that the structure of these materials is strongly affected by cooling rate. If quenched from above 800 °C in air, the entire composition line is single-phase. If cooled more slowly, the samples near the $LiNi_{0.5}Mn_{0.5}O_2$ end of the composition line phase separate into layered–layered nano-composites. If cooled even more slowly, these samples will form layered–layered composites or even layered–layered rocksalt materials. In terms of electrochemistry, the layered–layered composites were found to have lower capacity and a higher impedance than single-phase samples synthesized with the same metallic composition. This poor performance of the layered–layered composites may be attributed to the fact that one of the layered materials must have more nickel on the lithium layer than $LiNi_{0.5}Mn_{0.5}O_2$ (i.e., more than 10 %). When this nickel clusters during phase separation, lithium diffusion may be severely hindered. It is also of note that the material over which there has been the most debate, $Li_{1.2}Ni_{0.2}Mn_{0.6}O_2$ [44, 46], is very near to the boundary of the layered region when regular cooled. As such, differing opinions as to whether or not this material is single-phase may simply be due to slight differences in synthesis conditions and compositions.

VII Materials along the lithium-rich layered line from $LiNi_{0.5}Mn_{0.5}O_2$ to Li_2MnO_3 are generally considered to contain Ni^{2+} and Mn^{4+} [49, 50]. However, Simonin et al. [48] recently found that the magnetic moment in one such layered material was too low and this was attributed to some manganese being in the 3+ state.

This material, $Li_{1.2}Ni_{0.2}Mn_{0.6}O_2$, is in the "bump" region in the layered material on the Li–Mn–Ni–O face in Fig. 10.1. Materials in this part of the phase diagram were shown to contain metal site vacancies in Chap. 8. The presence of some vacancies could very well account for the lowering of the overall magnetic moment of the samples. X-ray absorption near-edge structure (XANES) and redox titrations confirmed that manganese was in fact in the 4+ state in these materials. This work also demonstrated that there is a solid solution line from Li_2MnO_3 to $Li[Ni_{1/6}\square_{1/6}Mn_{2/3}]O_2$. Along this line, all the transition metal layers have 2/3 manganese occupation allowing for ordering on two $\sqrt{3}$ x $\sqrt{3}$ superlattices. This gave rise to sharp superlattice peaks in the XRD patterns and the Monte Carlo simulation demonstrated that these structures do not phase separate into layered–layered composites.

VIII Some layered materials in the Li–Mn–Ni–O system lose a significant amount of lithium during synthesis. Typically, researchers counter this by adding a small amount of excess lithium and assume it is lost during synthesis (e.g. Refs. [51, 52]). However, the phase diagrams here show that most lithium-rich structures in the Li–Mn–Ni–O system can accommodate more lithium without phase separating. It can no longer be assumed that all excess lithium is lost during synthesis. Elemental analysis is therefore required to determine the composition of samples after heating. The contour plots generated in Chap. 7 can also be used to obtain approximate compositions. These contours show that as one adds excess lithium the *c* axis decreases rapidly, while the *a* axis is reduced more gradually. As such, it is vital to use both parameters in identifying final materials. Recognizing the actual compositions of the sample also helps explain differences in electrochemistry found in the literature. For example, Lin et al. [53] made a $LiNi_{0.5}Mn_{0.5}O_2$ sample without excess lithium that showed a lower capacity than materials made by Lu et al. [2]. The samples made by Lu used LiOH to cause coprecipitation of the mixed transition metal hydroxide precursor such that a small amount of excess lithium was inevitable. As such, the Lu sample was just within the layered region while the Lin sample was slightly outside. In the previous chapter, it was shown that this results in the single-phase material having a significantly higher capacity, consistent with the published results from Lu and Lin. A more extreme example of making $LiNi_{0.5}Mn_{0.5}O_2$ with excess lithium was published by Ohzuku et al. [52]. The contour plots were used to demonstrate that this sample lost little of the excess lithium during synthesis and the electrochemical performance was therefore attributed to being a lithium-rich layered material.

Two articles show very strange XRD patterns for $LiNi_{0.5}Mn_{0.5}O_2$.

IX The first was published by Jo et al. [54] and showed extreme peak broadening when regular cooled in oxygen. This was attributed to crystallites on the order of 14 nm. Here, it was shown that samples phase separate into layered-layered composites under such conditions. Therefore, the peak broadening seen by Jo was due to phase separation and not small crystallites since 14 nm is extremely small for having been heated for 3 h at 800 °C.

X The second strange result was obtained by Hinuma et al. [55] who found that a single-phase $LiNi_{0.5}Mn_{0.5}O_2$ underwent a complex phase separation upon annealing at 600 °C and a return to single-phase upon reheating to 1,000 °C. The new phases appearing upon annealing were not identified in the original paper. The transformations can now be understood. The layered boundary is high in the Gibbs triangle at high temperature such that the initial material synthesized at 1,000 °C was single-phase. Upon annealing at 600 °C the boundary was lower in the Gibbs triangle and the sample was then in the layered–layered rocksalt three-phase region determined in Chap. 6. The XRD pattern of the 600 °C annealed sample shown in the introduction is consistent with the R, N and M phases identified here. These transformations were also shown to be reversible in Chap. 7 explaining the return to single-phase upon re-heating to 1,000 °C.

XI Much work has been done in trying to develop layered-spinel composite electrodes in the Li–Mn–Ni–O system, but there was a poor understanding of the nature of the coexistence region. The work here shows that the layered-spinel region is small and ties to layered materials in the "bump" region discussed extensively in Chap. 8. Slow cooling in the layered-spinel two-phase region may result in the presence of some rocksalt material which grows rapidly as the three-phase regions transform. The rest of the coexistence region contains three-phase materials. Layered–layered–spinel materials can be made by quenching, though one of the layered materials is the disordered N phase discussed in Chap. 6. The N-material has poor electrochemistry and as such should be avoided. Kim et al. [94] have recently claimed to make a layered–layered spinel material under regular cooling conditions. This cannot be done as rocksalt will always be present in the three-phase samples that are not quenched. Instead the materials made by Kim were near enough to the Li–Mn line to be layered–spinel composites.

10.3 Future Work

There are now a wide variety of studies that can be performed using the PixSys solution-processing robot to synthesize combinatorial positive electrode materials. The first opportunity for continued research in combinatorial positive electrode materials is to adapt the combinatorial electrochemical method used by Fleischauer [95] for negative electrode materials. The primary challenge here is that the combinatorial positive electrodes are synthesized as powders and cannot be readily made by sputtering. The project of adapting the 64 channel combinatorial electrochemistry method for powders made using the PixSys robot is underway. Having both structural and electrochemical information for the combinatorial samples would be very useful in the search for promising new electrode materials.

Even without the electrochemical information, there are now a number of composition spaces that are worth studying with XRD of combinatorial samples. The first is to map out the single-phase regions in the Li–Co–Mn–Ni–O system. The main challenge with respect to mapping out the entire pseudo-quaternary system involves the coexistence regions. Since the layered boundaries are surfaces, it will be extremely difficult to determine directions of tie-lines. Nonetheless, it should be possible to identify approximate boundaries using the visual inspection of XRD patterns in a very large array of samples spanning hundreds of compositions. Mapping out the single-phase regions in the entire pyramid is therefore the main objective for on-going combinatorial work and this project has also already begun.

The robot can also be used to study any number of systems where a few single-phase structures of interest for positive electrode materials are already known. For example, $LiFePO_4$ has been used commercially as a relatively high power electrode material [96] and many efforts have been made to substitute a variety of atoms for either the iron or the phosphorous [97–99]. Efforts are now being made to develop a Li_2MSiO_4 material with M = (Fe, Mn) to cycle two lithium ions per formula unit [100]. All these efforts can be helped significantly by complete phase diagrams. Likely compositions where two-lithium cycling is possible could then be identified and tested.

The combinatorial approach discussed at length here for lithium containing layered oxides could also be of use in sodium-ion battery research where commercially viable electrode systems are still under development. Promising sodium layered oxide materials with varying amounts of iron and manganese on the transition metal layer are being studied [101, 102]. Making these materials with the combinatorial method is not trivial given that sodium loss during synthesis is more severe than lithium loss in layered materials. Assuming sodium loss can be reduced to a manageable level as was lithium loss in this thesis, a rapid screening of potential positive electrode materials for sodium-ion batteries could lead to new materials permitting sodium-ion batteries to become a viable alternative to lithium-ion batteries.

In terms of optimizing the search for promising positive electrode materials with the highest energy density, the current research strongly suggests that the best approach is to ensure that the samples lie within the single-phase layered region and in some cases they should be made as near as possible to the boundaries of these regions. The phase diagrams generated here should therefore be of significant value to Li-ion battery researchers. Two factors favoring the layered structures have been identified: increasing the synthesis temperature and lowering the oxygen partial pressure. Both of these conditions result in a larger single-phase layered region such that a greater range of compositions remain single-phase during regular cooling. There is therefore still a considerable amount of work that can be done in the Li–Mn–Ni–O pseudo-ternary system. Much of this requires synthesis of bulk samples to perform a variety of tests to find the optimum compositions and synthesis conditions for lithium-rich layered oxides.

There are also opportunities to make interesting core-shell materials in the Li–Mn–Ni–O system. One challenge with core-shell particles has always been to maintain the core-shell structure while heating at sufficiently high temperatures to make the desired materials. The quenched phase diagram shows the equilibrium phases such that tie-lines show which core-shell materials are possible during extended heating periods at that particular temperature. This therefore allows for careful selection of compositions and heating temperature to maintain the desired coexistence. This challenging project is also underway. Furthermore, as the combinatorial project in the Gibbs pyramid progresses, there will be opportunities to study new materials meeting the criteria listed in the introduction, namely high energy density and as little cobalt as possible to minimize cost.

References

1. J. R. Dahn and G. M. Ehrlich, *Handbook of Batteries, third edition, chapter 35*. New York: McGraw-Hill, 2002.
2. Z. Lu, L. Y. Beaulieu, R. A. Donaberger, C. L. Thomas, and J. R. Dahn "Synthesis, structure, and electrochemical behavior of $Li[Ni_xLi_{1/3-2x/3}Mn_{2/3-x/3}]O_2$," *Journal of The Electrochemical Society*, vol. 149, no. 6, pp. A778–A791, 2002.
3. H. Koga, L. Croguennec, P. Mannessiez, M. Ménétrier, F. Weill, L. Bourgeois, M. Duttine, E. Suard, and C. Delmas, "$Li_{1.20}Mn_{0.54}Co_{0.13}Ni_{0.13}O_2$ with different particle sizes as attractive positive electrode materials for lithium-ion batteries: insights into their structure," *The Journal of Physical Chemistry C*, vol. 116, no. 25, pp. 13497–13506, 2012.
4. Y. Cho, Y. S. Lee, S. A. Park, Y. Lee, and J. Cho, "$LiNi_{0.8}Co_{0.15}Al_{0.05}O_2$ cathode materials prepared by TiO_2 nanoparticle coatings on $Ni_{0.8}Co_{0.15}Al_{0.05}(OH)_2$ precursors," *Electrochimica Acta*, vol. 56, no. 1, pp. 333–339, 2010.
5. D. D. MacNeil, Z. Lu, and J. R. Dahn "Structure and electrochemistry of $Li[Ni_xCo_{1-2x}Mn_x]O_2$ $(0 \leq x \leq 1/2)$," *Journal of The Electrochemical Society*, vol. 149, no. 10, pp. A1332–A1336, 2002.
6. Y. Gao and J. R. Dahn, "The high temperature phase diagram of $Li_{1+x}Mn_{2-x}O_4$ and its implications," *Journal of The Electrochemical Society*, vol. 143, no. 6, pp. 1783–1788, 1996.
7. T. Takada, H. Hayakawa, E. Akiba, F. Izumi, and B. C. Chakoumakos "Novel synthesis process and structure refinements of $Li_4Mn_5O_{12}$ for rechargeable lithium batteries," *Journal of Power Sources*, vol. 68, no. 2, pp. 613–617, 1997.
8. J. M. Paulsen and J. R. Dahn, "Phase diagram of Li-Mn-O spinel in air," *Chemistry of Materials*, vol. 11, no. 11, pp. 3065–3079, 1999.
9. Z. Lu, D. D. MacNeil, and J. R. Dahn "Layered $Li[Ni_xCo_{1-2x}Mn_x]O_2$ cathode materials for lithium-ion batteries," *Electrochemical and Solid-State Letters*, vol. 4, no. 12, pp. A200–A203, 2001.
10. G. H. Carey and J. R. Dahn, "Combinatorial synthesis of mixed transition metal oxides for lithium-ion batteries," *ACS Combinatorial Science*, vol. 13, no. 2, pp. 186–189, 2011.
11. S.-H. Kang and K. Amine, "Synthesis and electrochemical properties of layer-structured 0.5 $Li(Ni_{0.5}Mn_{0.5})O_2$–0.5 $Li(Li_{1/3}Mn_{2/3})O_2$ solid mixture," *Journal of Power sources*, vol. 124, no. 2, pp. 533–537, 2003.
12. E. McCalla, A. W. Rowe, R. Shunmugasundaram, and J. R. Dahn "Structural study of the Li–Mn–Ni oxide pseudoternary system of interest for positive electrodes of Li-ion batteries," *Chemistry of Materials*, vol. 25, no. 6, pp. 989–999, 2013.
13. J. Cabana, S.-H. Kang, C. S. Johnson, M. M. Thackeray, and C. P. Grey "Structural and electrochemical characterization of composite layered-spinel electrodes containing Ni and Mn for Li-ion batteries," *Journal of the Electrochemical Society*, vol. 156, no. 9, pp. A730–A736, 2009.

E. McCalla, *Consequences of Combinatorial Studies of Positive Electrodes for Li-ion Batteries,* Springer Theses, DOI 10.1007/978-3-319-05849-8,

14. S.-H. Park, S.-H. Kang, C. S. Johnson, K. Amine, and M. M. Thackeray "Lithium–manganese–nickel-oxide electrodes with integrated layered–spinel structures for lithium batteries," *Electrochemistry Communications*, vol. 9, no. 2, pp. 262–268, 2007.
15. H. Koga, L. Croguennec, M. Ménétrier, P. Mannessiez, F. Weill, and C. Delmas "Different oxygen redox participation for bulk and surface: a possible global explanation for the cycling mechanism of $Li_{1.20}Mn_{0.54}Co_{0.13}Ni_{0.13}O_2$," *Journal of Power Sources*, 2013.
16. M. Sathiya, K. Ramesha, G. Rousse, D. Foix, D. Gonbeau, A. Prakash, M. Doublet, K. Hemalatha, and J. M. Tarascon "High performance $Li_2Ru_{1-y}Mn_yO_3$ ($0.2 \leq y \leq 0.8$) cathode materials for rechargeable lithium-ion batteries: Their understanding," *Chemistry of Materials*, vol. 25, no. 7, pp. 1121–1131, 2013.
17. A. R. Armstrong, M. Holzapfel, P. Novák, C. S. Johnson, S. H. Kang, M. M. Thackeray, and P. G. Bruce "Demonstrating oxygen loss and associated structural reorganization in the lithium battery cathode $Li[Ni_{0.2}Li_{0.2}Mn_{0.6}]O_2$," *Journal of the American Chemical Society*, vol. 128, no. 26, pp. 8694–8698, 2006.
18. J.-M. Kim, S. Tsuruta, and N. Kumagai, "Electrochemical properties of $Li(Li_{(1-x)/3}Co_xMn_{(2-2x)/3})O_2$ ($0 \leq x \leq 1$) solid solutions prepared by poly-vinyl alcohol (PVA) method," *Electrochemistry Communications*, vol. 9, no. 1, pp. 103–108, 2007.
19. H. Kawai, M. Nagata, H. Kageyama, H. Tukamoto, and A. R. West, "5 V lithium cathodes based on spinel solid solutions $Li_2Co_{1+X}Mn_{3-X}O_8$:$-1 \leq x \leq 1$," *Electrochimica Acta*, vol. 45, no. 1, pp. 315–327, 1999.
20. J. M. Amarilla, J. L. M. De Vidales, and R. M. Rojas "Electrochemical characteristics of cobalt-doped $LiCo_yMn_{2-y}O_4$ ($0 \leq y \leq 0.66$) spinels synthesized at low temperature from $Co_xMn_{3-x}O_4$ precursors," *Solid State Ionics*, vol. 127, no. 1–2, pp. 73–81, 2000.
21. L. Guohua, H. Ikuta, T. Uchida, and M. Wakihara "The spinel phases $LiM_yMn_{2-y}O_4$ (M = Co, Cr, Ni) as the cathode for rechargeable lithium batteries," *Journal of the Electrochemical Society*, vol. 143, no. 1, pp. 178–182, 1996.
22. J. G. Wen, J. Bareño, C. H. Lei, S. H. Kang, M. Balasubramanian, I. Petrov, and D. P. Abraham, "Analytical electron microscopy of $Li_{1.2}Co_{0.4}Mn_{0.4}O_2$ for lithium-ion batteries," *Solid State Ionics*, vol. 182, no. 1, pp. 98–107, 2011.
23. J. Bareño, M. Balasubramanian, S. H. Kang, J. G. Wen, C. H. Lei, S. V. Pol, I. Petrov, and D. P. Abraham, "Long-Range and local structure in the layered oxide $Li_{1.2}Co_{0.4}Mn_{0.4}O_2$," *Chemistry of Materials*, vol. 23, no. 8, pp. 2039–2050, 2011.
24. J. Bréger, M. Jiang, N. Dupré, Y. S. Meng, Y. Shao-Horn, G. Ceder, and C. P. Grey, "High-resolution X-ray diffraction, DIFFaX, NMR and first principles study of disorder in the Li_2MnO_3–$Li[Ni_{1/2}Mn_{1/2}]O_2$ solid solution," *Journal of Solid State Chemistry*, vol. 178, no. 9, pp. 2575–2585, 2005.
25. H. P. Klug and L. E. Alexander, *X-Ray Diffraction Procedures*. New York: John Wiley and Sons, 1974.
26. Y. S. Meng, Y. W. Wu, B. J. Hwang, Y. Li, and G. Ceder "Combining ab initio computation with experiments for designing new electrode materials for advanced lithium batteries: $LiNi_{1/3}Fe_{1/6}Co_{1/6}Mn_{1/3}O_2$," *Journal of The Electrochemical Society*, vol. 151, no. 8, pp. A1134–A1140, 2004.
27. S. Ping Ong, L. Wang, B. Kang, and G. Ceder, "Li- Fe- P- O_2 phase diagram from first principles calculations," *Chemistry of Materials*, vol. 20, no. 5, pp. 1798–1807, 2008.
28. A. Jain, G. Hautier, C. J. Moore, S. Ping Ong, C. C. Fischer, T. Mueller, K. A. Persson, and G. Ceder "A high-throughput infrastructure for density functional theory calculations," *Computational Materials Science*, vol. 50, no. 8, pp. 2295–2310, 2011.
29. S. P. Ong, A. Jain, G. Hautier, M. Kocher, S. Cholia, D. Gunter, D. Bailey, K. Skinner, K. Persson, and G. Ceder, *http://materialsproject.org*, 2011.
30. E. Antolini "Hexagonal-to-cubic phase transition by Li_2O evaporation from ordered $Li_xNi_{1-x}O$ solid solution," *Physica Status Solidi (A)*, vol. 173, no. 2, pp. 357–364, 1999.

31. T. Sata "High-temperature vaporization of Li_2O component from solid solution $Li_xNi_{1-x}O$ in air," *Ceramics International*, vol. 24, no. 1, pp. 53–59, 1998.
32. J. R. Dahn, U. von Sacken, and C. A. Michal "Structure and electrochemistry of $Li_{1\pm y}NiO_2$ and a new Li_2NiO_2 phase with the $Ni(OH)_2$ structure," *Solid State Ionics*, vol. 44, no. 1, pp. 87–97, 1990.
33. R. Moshtev, P. Zlatilova, V. Manev, and K. Tagawa "Synthesis of $LiNiO_2$ in air atmosphere: X-ray diffraction characterization and electrochemical investigation," *Journal of Power Sources*, vol. 62, no. 1, pp. 59–66, 1996.
34. Q. Zhong, A. Bonakdarpour, M. Zhang, Y. Gao, and J. R. Dahn "Synthesis and electrochemistry of $LiNi_xMn_{2-x}O_4$," *Journal of The Electrochemical Society*, vol. 144, no. 1, pp. 205–213, 1997.
35. H. Kobayashi, H. Sakaebe, H. Kageyama, K. Tatsumi, Y. Arachi, and T. Kamiyama "Changes in the structure and physical properties of the solid solution $LiNi_{1-x}Mn_xO_2$ with variation in its composition," *Journal of Materials Chemistry*, vol. 13, no. 3, pp. 590–595, 2003.
36. W. Li, J. N. Reimers, and J. R. Dahn "Crystal structure of $Li_xNi_{2-x}O_2$ and a lattice-gas model for the order-disorder transition," *Physical Review B*, vol. 46, no. 6, pp. 3236–3246, 1992.
37. L.-W. Ma, B.-Z. Chen, X.-C. Shi, W. Zhang, and X.-Y. Yang "Structure and stability of Li-Mn-Ni composite oxides as lithium ion sieve precursors in acidic medium," *Journal of Central South University of Technology*, vol. 18, pp. 314–318, 2011.
38. J. Cabana, M. Casas-Cabanas, F. O. Omenya, N. A. Chernova, D. Zeng, M. S. Whittingham, and C. P. Grey "Composition-structure relationships in the Li-ion battery electrode material $LiNi_{0.5}Mn_{1.5}O_4$," *Chemistry of Materials*, vol. 24, no. 15, pp. 2952–2964, 2012.
39. S. H. Park, S. W. Oh, S. H. Kang, I. Belharouak, K. Amine, and Y. K. Sun "Comparative study of different crystallographic structure of $LiNi_{0.5}Mn_{1.5}O_{4-\delta}$ cathodes with wide operation voltage (2.0-5.0 v)," *Electrochimica Acta*, vol. 52, no. 25, pp. 7226–7230, 2007.
40. Y. Idemoto, H. Narai, and N. Koura "Crystal structure and cathode performance dependence on oxygen content of $LiMn_{1.5}Ni_{0.5}O_4$ as a cathode material for secondary lithium batteries," *Journal of Power Sources*, vol. 119, pp. 125–129, 2003.
41. S. T. Myung, S. Komaba, N. Kumagai, H. Yashiro, H. T. Chung, and T. H. Cho "Nanocrystalline $LiNi_{0.5}Mn_{1.5}O_4$ synthesized by emulsion drying method," *Electrochimica Acta*, vol. 47, no. 15, pp. 2543–2549, 2002.
42. Y. Meng, G. Ceder, C. Grey, W.-S. Yoon, M. Jiang, J. Breger, and Y. Shao-Horn "Cation ordering in layered O3 $Li[Ni_xLi_{1/3-2x/3}Mn_{2/3-x/3}]O_2$ ($0 \leq x \leq 1/2$) compounds," *Chemistry of Materials*, vol. 17, no. 9, pp. 2386–2394, 2005.
43. M. M. Thackeray, C. S. Johnson, J. T. Vaughey, N. Li, and S. A. Hackney "Advances in manganese-oxide composite electrodes for lithium-ion batteries," *Journal of Materials Chemistry*, vol. 15, no. 23, pp. 2257–2267, 2005.
44. C. H. Lei, J. Bareño, J. G. Wen, I. Petrov, S. H. Kang, and D. P. Abraham, "Local structure and composition studies of $Li_{1.2}Ni_{0.2}Mn_{0.6}O_2$ by analytical electron microscopy," *Journal of Power Sources*, vol. 178, no. 1, pp. 422–433, 2008.
45. Z. Lu, Z. Chen, and J. R. Dahn "Lack of cation clustering in $Li[Ni_xLi_{1/3-2x/3}Mn_{2/3-x/3}]O_2$ ($0 \leq x \leq 1/2$) and $Li[Cr_xLi_{(1-x)/3}Mn_{(2-2x)/3}]O_2$ ($0 \leq x \leq 1$)," *Chemistry of Materials*, vol. 15, no. 16, pp. 3214–3220, 2003.
46. K. A. Jarvis, Z. Deng, L. F. Allard, A. Manthiram, and P. J. Ferreira "Atomic structure of a lithium-rich layered oxide material for lithium-ion batteries: Evidence of a solid solution," *Chemistry of Materials*, vol. 23, no. 16, pp. 3614–3621, 2011.
47. C. H. Lei, J. G. Wen, M. Sardela, J. Bareño, I. Petrov, S. H. Kang, and D. P. Abraham, "Structural study of Li_2MnO_3 by electron microscopy," *Journal of Materials Science*, vol. 44, no. 20, pp. 5579–5587, 2009.
48. L. Simonin, J.-F. Colin, V. Ranieri, E. Canévet, J.-F. Martin, C. Bourbon, C. Baehtz, P. Strobel, L. Daniel, and S. Patoux "In situ investigations of a Li-rich Mn–Ni layered oxide for Li-ion batteries," *Journal of Materials Chemistry*, vol. 22, no. 22, pp. 11316–11322, 2012.

49. S. Hy, W. N. Su, J. M. Chen, and B. J. Hwang "Soft x-ray absorption spectroscopic and raman studies on $Li_{1.2}Ni_{0.2}Mn_{0.6}O_2$ for lithium-ion batteries," *The Journal of Physical Chemistry C*, vol. 116, no. 48, pp. 25242–25247, 2012.
50. M. Jiang, B. Key, Y. S. Meng, and C. P. Grey "Electrochemical and structural study of the layered, Li-excess lithium-ion battery electrode material $Li[Li_{1/9}Ni_{1/3}Mn_{5/9}]O_2$," *Chemistry of Materials*, vol. 21, no. 13, pp. 2733–2745, 2009.
51. A. van Bommel, L. J. Krause, and J. R. Dahn "Investigation of the irreversible capacity loss in the lithium-rich oxide $Li[Li_{1/5}Ni_{1/5}Mn_{3/5}]O_2$," *Journal of The Electrochemical Society*, vol. 158, no. 6, pp. A731–A735, 2011.
52. T. Ohzuku, M. Nagayama, K. Tsuji, and K. Ariyoshi, "High-capacity lithium insertion materials of lithium nickel manganese oxides for advanced lithium-ion batteries: toward rechargeable capacity more than 300 mA hg^{-1}," *Journal of Materials Chemistry*, vol. 21, no. 27, pp. 10179–10188, 2011.
53. H. Lin, J. Zheng, and Y. Yang "The effects of quenching treatment and AlF_3 coating on $LiNi_{0.5}Mn_{0.5}O_2$ cathode materials for lithium-ion battery," *Materials Chemistry and Physics*, vol. 119, no. 3, pp. 519–523, 2010.
54. E. B. Jo, S. H. Ju, H. C. Jang, and Y. C. Kang "The properties of Li-Ni-Mn-O powders prepared by spray pyrolysis," *Journal of Ceramic Processing Research*, vol. 8, no. 5, pp. 352–355, 2007.
55. Y. Hinuma, Y. S. Meng, K. Kang, and G. Ceder, "Phase transitions in the $LiNi_{0.5}Mn_{0.5}O_2$ system with temperature," *Chemistry of Materials*, vol. 19, no. 7, pp. 1790–1800, 2007.
56. I. Belharouak, G. M. Koenig Jr, J. Ma, D. P. Wang, and K. Amine "Identification of $LiNi_{0.5}Mn_{1.5}O_4$ spinel in layered manganese enriched electrode materials," *Electrochemistry Communications*, vol. 13, no. 3, pp. 232–236, 2011.
57. H. Deng, I. Belharouak, R. E. Cook, H. Wu, Y.-K. Sun, and K. Amine "Nanostructured lithium nickel manganese oxides for lithium-ion batteries," *Journal of The Electrochemical Society*, vol. 157, no. 4, pp. A447–A452, 2010.
58. A. van Bommel and J. R. Dahn, "Analysis of the growth mechanism of coprecipitated spherical and dense nickel, manganese, and cobalt-containing hydroxides in the presence of aqueous ammonia," *Chemistry of Materials*, vol. 21, no. 8, pp. 1500–1503, 2009.
59. P. R. Bevington, *Data Reduction and Error Analysis for the Physical Sciences*. New York: McGraw Hill, 1969.
60. B. E. Warren, *X-Ray Diffraction, second edition*. New York: Dover Publications Inc., 1990.
61. S. K. Chatterjee and S. P. S. Gupta, "An integral breadth analysis for particle size and strain determinations in cold-worked fcc alloys," *Journal of Physics D: Applied Physics*, vol. 5, no. 3, pp. 609–612, 1972.
62. V. Berbenni, V. Massarotti, D. Capsoni, R. Riccardi, A. Marini, and E. Antolini "Structural and microstructural study of the formation of the solid solution $Li_xNi_{1-x}O$," *Solid State Ionics*, vol. 48, no. 1, pp. 101–111, 1991.
63. T. Marks, S. Trussler, A. J. Smith, D. Xiong, and J. R. Dahn "A guide to Li-ion coin-cell electrode making for academic researchers," *Journal of the Electrochemical Society*, vol. 158, no. 1, pp. A51–A57, 2011.
64. S. T. Myung, S. Komaba, K. Kurihara, K. Hosoya, N. Kumagai, Y. K. Sun, I. Nakai, M. Yonemura, and T. Kamiyama "Synthesis of $Li[(Ni_{0.5}Mn_{0.5})_{1-x}Li_x]O_2$ by emulsion drying method and impact of excess Li on structural and electrochemical properties," *Chemistry of Materials*, vol. 18, no. 6, pp. 1658–1666, 2006.
65. N. Metropolis, A. W. Rosenbluth, M. N. Rosenbluth, A. H. Teller, and E. Teller "Equation of state calculations by fast computing machines," *The Journal of Chemical Physics*, vol. 21, p. 1087, 1953.
66. E. McCalla, G. H. Carey, and J. R. Dahn "Lithium loss mechanisms during synthesis of layered $Li_xNi_{2-x}O_2$ for lithium ion batteries," *Solid State Ionics*, vol. 219, pp. 11–19, 2012.
67. G. R. Gavalas, S. Edelstein, M. Flytzani-Stephanopoulos, and T. A. Weston "Alkali-alumina sorbents for high-temperature removal of SO_2," *AIChE journal*, vol. 33, no. 2, pp. 258–266, 1987.

68. S. M. Lala, L. A. Montoro, and J. M. Rosolen, "$LiCoO_2$ sub-microns particles obtained from micro-precipitation in molten stearic acid," *Journal of Power Sources*, vol. 124, no. 1, pp. 118–123, 2003.
69. A. R. Naghash and J. Y. Lee, "Preparation of spinel lithium manganese oxide by aqueous co-precipitation," *Journal of Power Sources*, vol. 85, no. 2, pp. 284–293, 2000.
70. E. Antolini "Lithium loss kinetics from polycrystalline $Li_xNi_{1-x}O$ at high temperatures," *Journal of Materials Chemistry*, vol. 8, no. 12, pp. 2783–2786, 1998.
71. J. B. Goodenough, D. G. Wickham, and W. J. Croft "Some magnetic and crystallographic properties of the system $Li_x^+Ni_{1-2x}^{++}Ni_x^{+++}O$," *Journal of Physics and Chemistry of Solids*, vol. 5, no. 1, pp. 107–116, 1958.
72. R. Gunawan and D. Zhang, "Thermal stability and kinetics of decomposition of ammonium nitrate in the presence of pyrite," *Journal of Hazardous Materials*, vol. 165, no. 1, pp. 751–758, 2009.
73. R. D. Vold and G. S. Hattiangdi, "Characterization of heavy metal soaps by X-ray diffraction," *Industrial & Engineering Chemistry*, vol. 41, no. 10, pp. 2311–2320, 1949.
74. V. Berbenni and A. Marini, "Thermogravimetry and X-ray diffraction study of the thermal decomposition processes in Li_2CO_3–$MnCO_3$ mixtures," *Journal of Analytical and Applied Pyrolysis*, vol. 62, no. 1, pp. 45–62, 2002.
75. E. McCalla, C. M. Lowartz, C. R. Brown, and J. R. Dahn "Formation of layered-layered composites in the Li-Co-Mn oxide pseudoternary system during slow cooling," *Chemistry of Materials*, vol. 25, no. 6, pp. 912–918, 2013.
76. C. R. Brown, E. McCalla, and J. R. Dahn "Analysis of the cubic spinel region of the Li-Co-Mn oxide pseudo-ternary system," *accepted by Solid State Ionics*, 2013.
77. S. Levasseur, M. Menetrier, E. Suard, and C. Delmas "Evidence for structural defects in non-stoichiometric HT-$LiCoO_2$: electrochemical, electronic properties and ^7Li NMR studies," *Solid State Ionics*, vol. 128, no. 1, pp. 11–24, 2000.
78. Y. Sun, Y. Shiosaki, Y. Xia, and H. Noguchi "The preparation and electrochemical performance of solid solutions $LiCoO_2$-Li_2MnO_3 as cathode materials for lithium ion batteries," *Journal of Power Sources*, vol. 159, no. 2, pp. 1353–1359, 2006.
79. A. Van der Ven, M. K. Aydinol, G. Ceder, G. Kresse, and J. Hafner, "First-principles investigation of phase stability in Li_xCoO_2," *Physical Review B*, vol. 58, no. 6, p. 2975, 1998.
80. E. McCalla and J. R. Dahn, "The spinel and cubic rocksalt solid-solutions in the Li-Mn-Ni oxide pseudo-ternary system," *Solid State Ionics*, vol. 242, pp. 1–9, 2013.
81. J. S. Kasper and J. S. Prener, "The crystal structure of Mg_6MnO_8," *Acta Crystallographica*, vol. 7, no. 3, pp. 246–248, 1954.
82. G. W. Brindley "The effect of grain or particle size on X-ray reflections from mixed powders and alloys, considered in relation to the quantitative determination of crystalline substances by X-ray methods," *Philosophical Magazine*, vol. 36, no. 256, pp. 347–369, 1945.
83. B. J. Neudecker, R. A. Zuhr, B. S. Kwak, J. B. Bates, and J. D. Robertson "Lithium manganese nickel oxides $Li_x(Mn_yNi_{1-y})_{2-x}O_2$ i. synthesis and characterization of thin films and bulk phases," *Journal of the Electrochemical Society*, vol. 145, no. 12, pp. 4148–4159, 1998.
84. E. McCalla, A. W. Rowe, C. R. Brown, L. R. P. Hacquebard, and J. R. Dahn "How phase transformations during cooling affect Li-Mn-Ni-O positive electrodes in lithium ion batteries," *Journal of The Electrochemical Society*, vol. 160, no. 8, pp. A1134–A1138, 2013.
85. F. N. Rhines, *Phase Diagrams in Metallurgy*. New York: McGraw-Hill., 1956.
86. T. Ohzuku, and Y. Makimura, "Layered lithium insertion material of $LiNi_{1/2}Mn_{1/2}O_2$: a possible alternative to $LiCoO_2$ for advanced lithium-ion batteries," *Chemistry Letters*, vol. 30, no. 8, pp. 744–745, 2001.
87. C. R. Fell, D. Qian, K. J. Carroll, M. Chi, J. L. Jones, and Y. S. Meng "Correlation between oxygen vacancy, microstrain, and cation distribution in lithium-excess layered oxides during the first electrochemical cycle," *Chemistry of Materials*, vol. 25, no. 9, pp. 1621–1629, 2013.

88. Y. Cho, S. Lee, Y. Lee, T. Hong, and J. Cho "Spinel-layered core-shell cathode materials for Li-ion batteries," *Advanced Energy Materials*, vol. 1, no. 5, pp. 821–828, 2011.
89. E. McCalla, A. W. Rowe, J. Camardese, and J. R. Dahn "The role of metal site vacancies in promoting Li-Mn-Ni-O layered solid-solutions," *Chemistry of Materials*, vol. 25, no. 13, pp. 2716–2721, 2013.
90. C. S. Johnson, J.-S. Kim, A. J. Kropf, A. J. Kahaian, J. T. Vaughey, L. M. Fransson, K. Edström, and M. M. Thackeray "Structural characterization of layered $Li_xNi_{0.5}Mn_{0.5}O_2$ ($0 \leq x \leq 2$) oxide electrodes for Li batteries," *Chemistry of Materials*, vol. 15, no. 12, pp. 2313–2322, 2003.
91. M. Gu, I. Belharouak, J. Zheng, H. Wu, J. Xiao, A. Genc, K. Amine, S. Thevuthasan, D. R. Baer, and J. G. Zhang "Formation of the spinel phase in the layered composite cathode used in Li-ion batteries," *ACS Nano*, vol. 7, no. 1, pp. 760–767, 2012.
92. E. McCalla, J. Li, A. W. Rowe, and J. R. Dahn "The negative impact of layered-layered composites on the electrochemistry of Li-Mn-Ni-O positive electrodes for lithium-ion batteries," *Journal of The Electrochemical Society*, vol. 161, no. 4, pp. A606–A613, 2014.
93. T. Ohzuku and Y. Makimura, "Layered lithium insertion material of $LiCo_{1/3}Ni_{1/3}Mn_{1/3}O_2$ for lithium-ion batteries," *Chemistry Letters*, vol. 30, no. 7, pp. 642–643, 2001.
94. D. Kim, G. Sandi, J. R. Croy, K. G. Gallagher, S.-H. Kang, E. Lee, M. D. Slater, C. S. Johnson, and M. M. Thackeray "Composite 'layered-layered-spinel' cathode structures for lithium-ion batteries," *Journal of The Electrochemical Society*, vol. 160, no. 1, pp. A31–A38, 2013.
95. M. Fleischauer and J. Dahn, "Combinatorial investigations of the Si-Al-Mn system for Li-ion battery applications," *Journal of The Electrochemical Society*, vol. 151, no. 8, pp. A1216–A1221, 2004.
96. P. Marks "Dawn of motorsport's electric dream," *New Scientist*, vol. 211, no. 2831, pp. 26–27, 2011.
97. N. Jayaprakash, N. Kalaiselvi, and P. Periasamy "Synthesis and characterization of $LiM_XFe_{1-X}PO_4$ (M = Cu, Sn; X = 0.02) cathodes-a study on the effect of cation substitution in $LiFePO_4$ material," *International Journal of Electrochemical Sciences*, vol. 3, pp. 476–488, 2008.
98. H.-Y. Hu, W.-H. Qiu, F.-X. Li, H.-L. Zhao, and B.-Y. Wang "Influence of the Mg-substitution on electrochemical performances of $LiFePO_4$," *Chinese Journal of Power Sources*, vol. 30, no. 1, p. 18, 2006.
99. J. Hong, C. S. Wang, X. Chen, S. Upreti, and M. S. Whittingham "Vanadium modified $LiFePO_4$ cathode for Li-ion batteries," *Electrochemical and Solid-State Letters*, vol. 12, no. 2, pp. A33–A38, 2009.
100. D. Rangappa, K. D. Murukanahally, T. Tomai, A. Unemoto, and I. Honma, "Ultrathin nanosheets of Li_2MSiO_4 (M = Fe, Mn) as high-capacity Li-ion battery electrode," *Nano Letters*, vol. 12, no. 3, pp. 1146–1151, 2012.
101. N. Yabuuchi, M. Kajiyama, J. Iwatate, H. Nishikawa, S. Hitomi, R. Okuyama, R. Usui, Y. Yamada, and S. Komaba "P2-type Na_x $[Fe_{1/2}Mn_{1/2}]O_2$ made from earth-abundant elements for rechargeable na batteries," *Nature Materials*, vol. 11, no. 6, pp. 512–517, 2012.
102. J. Thorne, R. Dunlap, and M. Obrovac "Structure and electrochemistry of $Na_xFe_xMn_{1-x}O_2$ ($1.0 \leq x \leq 0.5$) for Na-ion battery positive electrodes," *Journal of The Electrochemical Society*, vol. 160, no. 2, pp. A361–A367, 2013.

Index

E. McCalla, *Consequences of Combinatorial Studies of Positive Electrodes for Li-ion Batteries,* Springer Theses, DOI 10.1007/978-3-319-05849-8,

MIX
Papier aus verantwortungsvollen Quellen
Paper from responsible sources
FSC® C105338

If you have any concerns about our products,
you can contact us on
ProductSafety@springernature.com

In case Publisher is established outside the EU,
the EU authorized representative is:
Springer Nature Customer Service Center GmbH
Europaplatz 3, 69115 Heidelberg, Germany

Printed by Libri Plureos GmbH
in Hamburg, Germany